The New Guide to Ramsgate, Margate, Broadstairs, and St. Peter's, containing a history of the Isle of Thanet, and lifeboat work on the coast of Kent. With map of Thanet, etc.

Anonymous

The New Guide to Ramsgate, Margate, Broadstairs, and St. Peter's, containing a history of the Isle of Thanet, and lifeboat work on the coast of Kent. With map of Thanet, etc.
Anonymous
British Library, Historical Print Editions
British Library
1886
98 p. ; 8º.

THE

NEW GUIDE

TO

RAMSGATE, MARGATE, BROADSTAIRS,

AND

ST. PETER'S,

CONTAINING A

HISTORY of the ISLE of THANET,

AND

LIFEBOAT WORK

ON THE COAST OF KENT.

WITH MAP OF THANET,

¾ INCH SCALE.

RAMSGATE:
PRINTED AND PUBLISHED BY S. R. WILSON,
36, HARBOUR STREET.

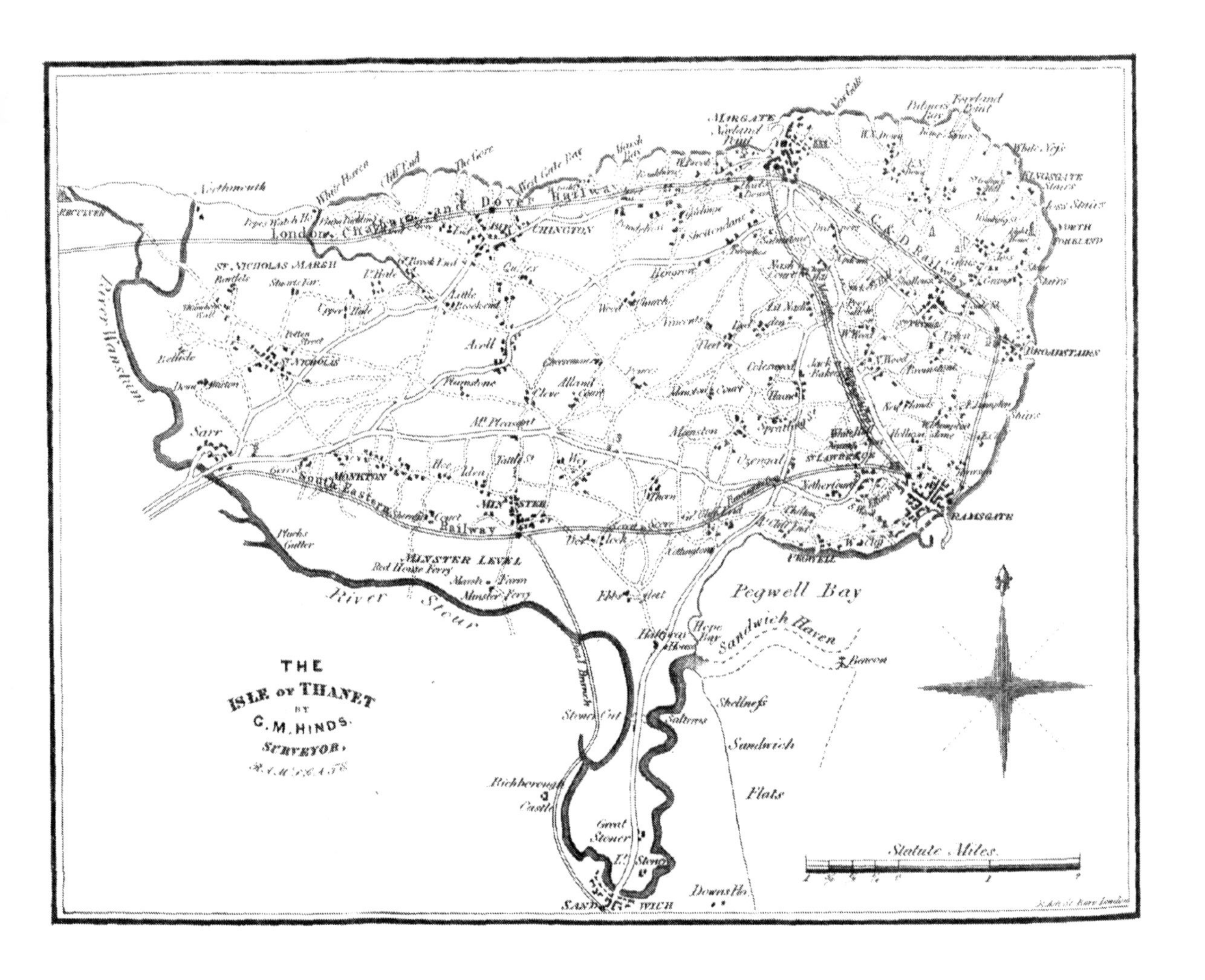
THE
ISLE OF THANET
BY
C. M. HINDS.
SURVEYOR.
MARGATE
RAMSGATE
BROADSTAIRS
MINSTER
MONKTON
ST. NICHOLAS
ST. NICHOLAS MARSH
MINSTER LEVEL
Pegwell Bay
Sandwich Haven
Sandwich
Flats
River Stour
River Wantsum
Statute Miles.
Sarr
Richborough Castle
Great Stoner
Beacon
London Chatham and Dover Railway
South Eastern Railway
Mt. Pleasant
Reculver
North Foreland
Kingsgate
Pegwell
Sandwich

Isle of Thanet.

THE ISLE OF THANET has been said by most writers to be the Inis (an Island) Ruim or Ruochim of the Britons, that is, the Island of Richborough ; but as Richborough in early times was an Island, this is rather doubtful. The earliest mention of it as anything approaching to its present appellation is by Junius Solinus, a Roman writer, from whom the idea springs that the name of Thanet is derived from the Greek *athanatos* (immortal) ; he speaks of it as Athanaton and Thanaton. By some writers it is ascribed to Isidore of Seville, as Thanatos, (Death) but this like so many of the haphazard speculations of a recent compiler of Guides, will not bear scrutiny ; had he exercised any amount of research, or even had he considered, (if he knew it), the time when the said Isidore wrote, viz., about 620 to 630, it would have been quite sufficient to settle such an idea, as at that time the Danes had not commenced their ravages, and instead of being the land of Death, it was the land of Life and Peace, which, with the exception of the Danish incursions, it has continued to be, up to the present time. Will such wild speculators take the trouble to ask where crops are heavier, corn better, and where has God given a greater increase than in this Island of Thanet? It was afterwards called Teneth and Tenetlonde by the Saxons, from which it has now liquified into Thanet Land or Isle of Thanet. Another probable idea is that Thanet is derived from a Saxon word *tene*, a fire or beacon, from the great number of watch-fires kept up on this important outpost to warn the neighbouring country when strange sails were observed, as was very needful in those disturbed times when the invasions of the Danes were so frequent. It was visited by these Pirates on account of its propinquity to the port of Richborough, (at that time a very important town.) The Island appears to have been separated on the south and west sides from the adjacent lands, by a broad estuary, the course of which is now marked on the south side by the river Stour, which, after a very circuitous route by Sandwich empties itself into Pegwell Bay; and on the west by a small stream originally called Genlade, now Wantsume or Wentsome, which, running past Sarre, flowed into the sea at Northmouth, but now, from its dispersion among the land, is hardly a continuous stream. It has been endeavoured to make out, that Cantium derived its name from the point at Deal, at that time jutting or *canting* out into the Sea, but such shallow reasoning will not hold water, as this was undoubtedly the Latin name given to it by the Roman Invaders and would not be likely to be made up from the Anglo-Saxon Cant; and again the jutting out into the Sea would not be so apparent at that time when undoubtedly the coast of Thanet extended out as far as the Goodwin Sands, and the Straits of Dover had only just been broken through, being originally an Isthmus.

We know from the writings of the ancient Chroniclers that this estuary was in early times navigable, and that, by ships of considerable size, for we find, in 1052, King Harold's fleet sailed round the Island and plundered the east coast of Kent, going out to sea by Northmouth. Again, Earl Godwin took his revenge for being outlawed, by ravaging this coast, plundering the maritime towns, sailing into the harbour of Sandwich, and thence to Northmouth and London. Irvine in his treatise, de rebus Albionicis, says, "Erat olim in hoc fluvio statio primissima navibus et gratissima nautis Sarra nominata." Leland, who wrote in Henry VIII.'s time, says, in his Itinerary, "At Northmouth, where ye estery of ye se was, ye salt water swellith yet up at a Creeke a myle and more, toward a place called Sarre, which was the commune ferry when Thanet was fulle iled." Another and strong proof that vessels, sailing between France and London, found this inlet useful, is the fact, on the authority of Hasted, that the town of Sandwich was originally termed Lundenivic, from its being the entrance to the Port of London; he says, "for so it was on the sea coast, and it retained its name until the supplanting of the Saxons by the Danes." And, from the same source, we find, where "Sandwich now stands is supposed, in the time of the Romans and before the decay of the haven, Portus Rutupinus (Richborough) to have been covered with that water, which formed the bay of it, which was so large that it is said to have extended far beyond this place, on the one side, almost to Ramsgate cliffs, and on the other, nearly five miles in width, over the whole flat of the sand on which Stonar, and Sandwich too, were afterwards built, and extending from thence to the estuary which then flowed up between the Isle of Thanet and the main land of this County." The two Roman Fortresses, erected at the opposite extremities of this passage, namely, Richborough and Reculver, appear also as strong evidence that this course was employed by vessels trading between London and the Continent, and that they were intended to guard the inlet. We have the testimony of an unexceptionable author, that through this port, lay the direct and accustomed passage to London by sea, beyond the middle of the fourth century; the author above referred to says: "It remained in its natural and perfect state, so long as the Romans enjoyed Britain, and no doubt for some time after. But in Bede's time, and perhaps an age before that, the port began to decline, by diminishing its breadth, for he tells us it was then but three furlongs wide, fordable in two places, and was called Wantsume, or the deficient water. There was, however, a considerable passage for ships, till about the time of the Norman Conquest, very soon after which great event, and change, the inhabitants, both on the Island and the mainland of Britain, began to dyke out the sea, and reclaim the land that formerly had been under water; and so those great changes, which we see now, in the present day, were caused by the sea, usually irresistible, and which generally encroaches on the land, which it surrounds, acted here re-reversely as through the sandy nature of the soil, the tides lost their strength, and the inhabitants recovered much land that before had been deep down under the sea. The stream that originally ran into the arm of the North Sea, and divided Thanet from the Continent, runs now, which shews, in some measure, the breadth of the old channel, a mile and a half east of Reculver, while the Stour makes its way into the sea at Sandwich, so that the Isle of Thanet, which was formerly separated from the Continent by the entire channel, or the old Portus Rhutupinus, or Ritupensis, and was then, as in its natural state, all high land, is now a peninsula, or at best a river isle only, with the Stour-Wantsume on the south, the Mile stream on the south-west, and the Nethergong Wantsume on the west. The rest of the island looks to the East and North Seas as heretofore; but the figure is altered from a circular, to an irregular oval, which circumstance is a very strong confirmation of the reality of the views here advanced." Considering the vessels of that date, and the time the trip would take, this route must have been of great advantage in shortening the passage, as well as in those days, when pirates and marauders of all sorts abounded on these inland seas, giving them some little security through part of their dangerous voyages from the mainland of Britain to the Continent. Venerable Bede, in describing the Island, as it was in his own time, says, "The Island of Tanet is of considerable bigness, (that is, according to the English way of reckoning, consisting of six hundred families,) which the river Wantsume divides from the Continent, which is about three furlongs broad, and passable over only in two places, namely Sarre and Sandwich, and goes into the sea at both its heads." And Lewis, commenting upon the above, and supposing each family, or hide of land, to consist of no more than 64 acres, which he deems would be the very lowest reckoning, says, "at that rate there will then have been 38,400 acres of land here, which is above double the number we have now without the marshes." Besides this, he also says, "This Island seems anciently to have been as large or larger than it is now; notwithstanding the addition of Stonar to it, and the inning so much land in the marshes, it is very plain that formerly the land on the north and east sides of it, from Westgate to Cliffsend, went much farther out unto the ocean than it does now." It is quite natural from the above testimony to suppose that previous to what we have any account of, the land somewhere between Broadstairs and Cliffsend, if not the greater part of the way, extended out into the sea, in the direction of the Goodwin Sands, and as there is every reason to conclude that these Sands were once dry land, there would be a very light draught of water between them and this Island, so that large vessels would have to sail at the back of the Sands, making the distance to the port of London greater.

With regard to the formation of this estuary, we gather from ancient writers, that the North Sea being stopped in its current by the Isthmus "which once was, where now is the pass between Dover and Calais, insinuated itself at the low grounds at the North Mouth or Yenlad, and by Ebbsfleete, running up those levels, to Sturmuth, Wingham, &c., and by Fordwich to Canterbury." But in process of time, by the continual beating of the sea, that Isthmus was worn away, hence the gradual subsiding of the water on these lands, and about the same time, the overflowing of the low Countries of Flanders and Holland. Lewis says, "although these lands were a kind of flat or shallow, yet until so much of the water found an outlet in the neighbouring parts of Flanders and the low Countries just over against them, (where it took up more than the space of 30 miles of the shore,) they were always overflowed; but from thenceforth, for want of that store of water which formerly overlaid them, they became a kind of dry land, which being interbanked and kept from the high water and spring tides, in process of time grew dry."

The Britons were the ancient inhabitants of this little Island of Thanet, and they were succeeded by the Romans. Coins of both these nations have been found; those of the Romans are two, namely, one of Constantine and one of Domitian; no coins whatever have been discovered, of the light haired Saxons, who drove out the native Britons, some little time after the latter had been abandoned by the Romans. But this Island was allotted to Hengist and Horsa, by the Britons, who sent for them on account of their being harassed and troubled by the Picts and Scots. These Saxon leaders (whose existence has been doubted) landed at Ebbs Fleet, the usual landing at that time, in or about the year 449. However, they were not content with this fertile piece of land, and very soon, being joined by reinforcements of Saxon troops, they overran the neighbouring land, and made themselves masters of a much more extensive domain, the chief seat of which was established at the Roman city of Durobernum, to which the Saxons gave the name of Cantwara-byrig, or the city of the Kentish men, which it still retains under the slightly altered form of Canterbury. Hengist thus became king of Kent, while King Vortigern was obliged to retire into Wales. Horsa was slain in one of the many battles with the Welsh in Kent. The consequence of these great national events was that the ancient inhabitants were miserably oppressed, their Christian Churches turned into Pagan Temples, and, to shew the completeness of their conquest, nearly every place altered in name to such as in their own language was intelligible. But there was a much greater misfortune to which the inhabitants of Thanet were subject; for, from its peculiarly exposed situation, they were constantly being visited by those merciless pirates, the Danes. In King Sigibert's reign, which commenced A.D. 635, we learn that they used to land here nearly every year and ravage and plunder the country and inhabitants. In 851 a great army of the Danes having taken up their Winter quarters here, for the first time, was defeated at Sandwich by King Athelstane and Duke Ealchere; and, during the same year, 60 of their ships were taken. Two years afterwards, in 853, they again landed a large force of men here, and in this invasion they were more successful, for they defeated the Dukes Ealchere and Wada, who had both the Canterbury and Sturry men under their command.

In the autumn of 865, during the reign of Ethelbert, the Danes again took up their quarters here and ravaged Thanet; they also made a league with the men of Kent, which, however, like all their other treaties, they soon broke. There were also invasions by the Danes in 980, 981 and 982, but not again till 988, when they burnt the Abbey at Minster, built by Dompneva, who was the first Abbess, with all the nuns, clergy and people who had taken refuge there.

The old proverb, "evil communications corrupt good manners," appears to have been exemplified in the case of the inhabitants, who were very soon found carrying on the same piracy, for, in 969, King Edgar ordered the Island to be plundered, because the Inhabitants had robbed some English merchants as they sailed through the Wantsume. Ethelred II. levied an army to give the Danes battle in 1002, and some of them fled to the Isle of Thanet, where he could not follow them. They again landed here in 1009. Two years after this, in 1011, the Danes, under Sweyn, King of Denmark, entirely destroyed the Monastery at Minster; they also put to the sword its nuns and Abbess. King Cnut bestowed its site and endowments upon the monks of St. Augustine. We hear nothing more after this of the barbarisms of the Danes; but in later years, when the town and port of Sandwich became so formidable to the French, the Island was constantly in fear of being invaded. It was here that Eustace le Moyne, the French Admiral, conducted Louis, the Dauphin of France, invited to be King of England, in 1216, to join the Barons, who, on account of the cruelty and fury of the savage King John, felt that any change, even that of a new conquest and French dynasty, was most desirable. After the death of John, the Barons thought rather different, and accordingly, Louis, having been defeated by them, had to flee; on his way back to France burning the town of Sandwich. Measures were not taken for the security of this Island until the 43rd year of the reign of Edward III., namely, the year 1369, when the King ordered John de Cobeham and others to look after the enclosing and fortifying of any such places in the Isle of Thanet, where ships and boats could land, with mounds and ditches; and also, that those whose land would be benefited by these fortifications, should be charged with the expense of such fortifications. It cannot be out of place, now that the Volunteers are an established force, to mention the fact that, during the war with France, upon the alarm of an invasion, a troop of Cavalry, six companies of Infantry, and three of Artillery bravely volunteered and enrolled themselves here, to repel the threatened attacks of an ambitious and envious foe; and no higher encomiums were paid by Reviewing Generals to any body of Volunteers in the Kingdom for their military discipline and soldier-like appearance. No doubt honourable mention will be again made of the Corps formed in the Isle of Thanet, and in case of an emergency, there is little fear but that the same gallant spirit will be displayed as heretofore.

The amount of historic interest with which the Island and its neighbourhood is invested is very great, but those who look for a small guide to give them all the information necessary, will be deceived. We must refer those readers who may feel an interest in this important, though well-nigh forgotten Island, to Hasted's Kent, Lewis's Antiquities of Thanet, Boys' Sandwich, Stanley's Canterbury, Hudson's Queen Bertha, and other more ancient writers. Here on this coast landed those who came to bring us our civilization; those who gave us our present national character, of which we now see the advantage, and reap the benefits; and those who, if not the first to plant Christianity in this country, at any rate were first to establish that root firmly; and of what three things are Britons more justly proud, than their civilization, their national character, and their Christianity.

In the first instance, that this was the landing place of our great civilizers.—It is generally supposed that Julius Cæsar landed somewhere near Deal, but there has been great controversy on this point, yet although it has been endeavoured to be disproved, the arguments at present seem strong as ever in its favour; and while the inhabitants of Cantium continually managed to worst him, so that his adversaries can say that he was not our civilizer, yet to him must be awarded the originality of conceiving that great idea, and to him we must of necessity give thanks.

for opening up our country to the civilized world, and giving us a glimpse of that light which we now see shining so brightly on all sides, even if to other we give the credit of carrying out and, as it were, finishing his design.

Secondly.—At Ebbs Fleet landed Hengist and Horsa, our great English Forefathers, the founders as it were of our present characteristics : steady indomitable perseverance a determination not lightly made, but when made rarely departed from. Stanley says of them—Their arrival was "one of the five great landings in English History which gave us our English forefathers and our English character"

There has lately been a good deal of controversy as to the existence of Hengist and Horsa. "These Worthies," Mr. Wright says, "appear to have belonged rather to the mythic poetry of the heroic ages of the north than to the sober annals of Saxon warfare in our island." The names are nearly synonymous in meaning, each signifying a horse, the animal reverenced by the Saxon, and in this sense the settlers in the Isle of Thanet may appropriately be called "the followers of Hengist and of Horsa." But if we consider the matter in this light, what must we do with the records of such authorities as Bede, Leland, Lewis and others? In fact, the existence of Hengist and Horsa, as human beings, seems so intimately connected with the History of England, that, to deny them as such, is to deny the authenticity of all traditional history. Far from their synonymous names being a reason for denying their existence it seems a very sufficient proof to the contrary, for, being brothers, what more likely than that they should be called by a name implying the same, but in sound very different. And undoubtedly we have from this the meaning of the rampant horse, as the seal of Kent, for, from these two brothers must this have taken its origin. We must quote the recorded facts before leaving this interesting matter, that Horsa was slain in battle at Aylesford, 455, and Hengist died after reigning 31 years as King of Kent.

Hudson's Life of Queen Bertha says:—

"In this emergency the rulers of Britain agreed to Vortigern's proposal, that the aid of the Saxons should be sought. Messengers were despatched without delay and on the ocean they fell in with three Saxon vessels cruising in search of plunder. Hengist, the commander of the piratical band, instantly steered for the coast of Kent, and they landed at Ebbs Fleet, in the Isle of Thanet, not far from the ancient town of Sandwich. Vortigern and others equally shortsighted, were delighted to welcome their allies sooner than they had expected them."

"Hengist was accompanied by Horsa his brother; they were high born Jutes, for their great grandfather Wecta was a son of Woden. Roger of Wendover gives their genealogy, and relates a conversation which took place when the Saxon chiefs stood before Vortigern."

"The King questioned them concerning their religion, they explained their creed and told him that their forefathers had dedicated the fourth day of the week to their great ancestor Woden, the sixth to Frea his wife, and the other days to the sun and moon, and to heroes who had lived in ancient time, "and who now govern the world" said the Saxon. "I grieve much" replied Vortigern, "for your belief or rather for your unbelief; but I am exceedingly rejoiced at your coming, which is most opportune for my urgent necessities. For I am pressed by my enemies on every side, and if ye will share with me the toil of fighting, ye shall remain in my kingdom, where ye shall be had in honour and enriched with lands and possessions." Old Roger goes on to tell us that the King collected his troops, and crossed the Humber to meet the Picts and Scots, and that there was not much need for the Britons to fight, because the Saxons fought so manfully that they immediately routed the enemy, who before their arrival had become habituated to conquest."

"Having gained so great a victory by their aid, Vortigern was lavish in gifts and promises. The Saxon was acute enough to see how he could take advantage of the King's gratitude; he seems also to have perceived that the British began to suspect the fidelity of their allies, and to blame their monarch's unguarded conduct; for he thus addressed Vortigern; "My Lord, thy enemies vex thee on all sides, and say they will depose thee, and bring Aurelius Ambrosius from Armorica, and make him King in thy room. If it please thee, therefore, let us send into our country, and invite over more soldiers that our number may be increased." The King agreed to the proposal, and bade him send into Germany for speedy aid. Straightway messengers were despatched, who brought back with them eighteen vessels, full of chosen soldiers. They also brought over Hengist's daughter, Rowena, whose bright beauty and fascinating manners, seduced Vortigern into committing even greater imprudence than he had hitherto been guilty of. His subjects grew discontented and rebellious, and he became harassed by difficulties with which his mind was unequal to cope."

"The old chroniclers paint Vortigern's character in very dark colours; they are indignant with him for not following the advice of his counsellors, and for not heeding the warning of Merlin, the venerated prophet of that day. Old

Roger gravely relates a circumstance, which ought to have made a deep impression on the mind of the British King; 'King Vortigern was sitting by the bank of a pool, out of which two dragons came forth; one of them was white, the other red. As soon as they approached each other, they commenced a dreadful combat, breathing forth flames. The white dragon got the better of his antagonist and pursued him into the margin of the pool; when the red one, indignant at the repulse, turned on the white dragon, and forced him to retire.' While they were thus fighting, the King commanded Merlin Ambrosius to say what the battle ment. Whereupon busting into tears, and full of the spirit of prophecy, he thus began:—'Woe to the red dragon, for his banishment approaches; the white dragon which signifies the Saxons whom thou hast invited over, shall possess his cavern; whereas the red dragon signifies the British people, which shall be opposed by the white dragon. His mountains shall be brought low as the valleys, and the rivers of the valleys shall flow with blood; the isles of the ocean shall be subdued by his might, his praise shall be sounded among the nations.'"

"But Rowena's words were more persuasive than the fiery breathings of the dragons. She had a winning way of blending the Saxon and the Celtic tongues, and her blue eye and sprightly wit had more power over the infatuated Vortigern than had the stern eye and wild rhapsodies of Merlin. Yet it could have needed no prophetic inspiration to foresee that awful calamities were impending; that two thunder clouds were meeting—the anarchy wrought by civil discord and the misery brought on by hostile invasion.

"Vortigern married Rowena, and thereby deeply offended his sons and his nobles. Hengist, his father-in-law soon presumed on the relationship, and found opportunity to speak to him thus :—'Listen to my counsel. Let us invite over my son Octa, with his brother Abissa, both warlike men, and give them the country in the north part of Britain, in the neighbourhood of the wall between Deira and Scotland. They will there sustain the attacks of the barbarians, and thou wilt remain in peace on this side of the Humber.'"

"Vortigern assented, and straightway on his invitation came Octa, Abissa, and Cerdic, with 300 vessels full of armed men; all of whom were graciously received by Vortigern and presented with rich gifts. The Britons, seeing this, and fearing their treachery, advised the King to expel them from his dominions, for pagans ought not to have intercourse with Christians. Besides so great a multitude had come over, that they were a terror to the natives whom they ought to have protected. But Vortigern declined their counsel, because he loved the Saxons above all people, for his wife's sake. The exasperated Britons revolted, and deprived the King, whom they despised, of his dignity and authority, which they conferred on Vortimer, the eldest of his three sons. The deposed Monarch seems to have spent the remainder of his life most ingloriously, contentedly yielding to the stronger will of his imperious father-in-law."

"We have authentic accounts of five battles fought in Kent. The first took place on the banks of the little river Derwent. The second at Eaglesford, now Aylesford, on the Medway, was a desperate struggle which lasted several days. Both armies fought long and fiercely, till the battle turned against the Saxons who fled from the field. The Britons pursued them closely, slew an immense number, and dispersed the remainder; but Horsa rallied a band of men, and attacked a body of Britons commanded by Catigern, a brother of King Vortimer, whom he slew, whereupon Vortimer pursued Horsa and killed him with his own hand. Thus each party lost a brave captain on that day, and Hengist is said to have fled from the enemy, a thing which he had never done before. (A.D. 455)

"A third battle took place either at Stonar, near Sandwich, or at Stone End, near Hythe; the similarity of these names has occasioned confusion, but in either case it was near the sea; which agrees with the account, that the Saxons, being repulsed, fled to their boats. An extraordinary number of human bones are still existing, crumbling away under the church at Hythe, where they are looked on with unavailing curiosity, as nothing certain is known about them: but it is supposed that some great battle must have occurred in the neighbourhood, and probably, it was that of Stone End.

"Hengist gained his crowning victory at Crayford, A.D. 457, after which he assumed the title of King of Kent.

"But his supremacy was that of a military despot; and the country was in such an unsettled state, that many historians distinguish Hengist as the conqueror, and his son Eric as the founder of the Kingdom.

"The last battle was fought in the Isle of Thanet, very near the spot on which Hengist and Horsa had first landed. Twelve British chieftains were left dead on the field, which was thenceforth called Wypped's Floet, after Wypped, the only Saxon thane who fell. The Britons are described as fleeing from the Saxons as from fire, after that final conflict.

"We have no reliable account of the latter years of Hengist's life; his death is supposed to have occurred about thirty years after that memorable day on which he and his brother, while cruising in the Channel, received Vortigern's invitation.

"It should be borne in mind that Hengist, the Saxon hero of the East, flourished shortly before King Arthur, the British hero of the West. Exaggeration has been equally busy with both these names. Hengist is said to have carried terror and devastation to the remotest corners of the island; but Turner, who gives Anglo-Saxon history very clearly, considers that there is no good evidence of his having penetrated, except in his first depredations, beyond the region which he transmitted to his posterity, viz., the present county of Kent, with Middlesex, and a part of Surrey."

Thirdly.—At the same place that saw the landing of Hengist and Horsa, namely Ebbs Fleet, nearly a century and a half after their disembarking on our shores, St. Augustine, the father of Christianity in Britain, first set foot on English ground. Surely, in wandering near this place, a strong feeling of thankfulness and love must fill our hearts to the men who tore away from overshadowed Britain, the dark cloud of heathenism, and who planted in our well-loved Island, the first small seeds of Christianity and civilization.

The man of the world, pushing forward, making for himself money and a name, strong minded and self willed, passes a cottage; his stern face relaxes, a smile is upon it, and thankfulness on his lips; he is saying, God bless her, the old dame, who, in that house, read to him the tale of Whittington and his Cat, and whispering in his ear Excelsior, told him that he might be a great man some day. The Christian too pauses on the threshold of a similar door, to bless one who taught him to look upwards, fear God, and love his fellow man. Shall we not then, in like manner, pause on the landmarks to be found in this Island, and, looking back, thank those who brought us our two greatest blessings?

In speaking of the commencement of Christianity in Britain, our starting point should be, the landing of Augustine at Ebbs Fleet; but my readers must bear with me and first come away to Rome, and see there, what prompted this great missionary undertaking, in which so much of our country's history is mixed, so that this subject may lose nothing of the importance it deserves. Gregory the Great, at the time we must first bring him to our mind, about 574, had not risen to the supremacy of St. Peter's chair, and was only a monk in St. Andrew's Monastery, but he was remarkable for his beneficence and kindness to the poor. Among interesting legends told of this good man, there is one which narrates that, when, according to his usual custom, he was engaged in feeding twelve beggars, at his marble table, there came another, who, the story tells, was an angel. It is also related that on one occasion, he inflicted severe penance upon himself, when he heard that a poor man had been starved to death, as if he had been answerable for it. It is not strange, when we look back at the many good deeds that are recorded of him that he should, in his wanderings about Rome, find his way to the market-place, where slaves of every land under the sun, were bought and sold, like beasts of the field, whose treatment, in many cases, was far superior to their own, and it will not require a great stretch of the imagination to suppose, that he picked out from the motley crowd, made up of Africans, Egyptians, and other dark skinned and gipsy-looking faces, three beautiful boys who had just been brought by a slave-merchant, boys with a fair skin and long fair hair, as English children would then have. He asked the merchant who owned them, from what part of the world they came, and whether they were Pagans or Christians. He was told that they were heathen boys from the far off Island of Britain; when Gregory heard this, he said, "Alas! more is the pity, that faces so full of light and brightness, should be in the hands of the Prince of Darkness; that such grace of outward appearance, should accompany minds without the grace of God within!" After making other inquiries he again asked the name of their nation, "Angles," he was told: "Angles," said Gregory; "they have the faces of *Angels*, and they ought to be made fellow-heirs of the Angels in heaven. But of what tribe or province of the Angles are they?" "Of Deira," said the merchant. "*Deira?*" exclaimed Gregory, "then they must be delivered from the wrath"—in Latin *de irâ*—"of God. And what is the name of their King?" "Ælla," he was answered. "*Ælla;* then *Alleluia* shall be sung in his land." We do not know whether the good and pious monk, Gregory, took these poor boys and educated them, which it seems likely he should wish to do, but he certainly made up his mind that he would organize a missionary enterprise to convert Britain, the land in which the youths had such bright and open faces as the three beautiful boys before mentioned. Even at that remote period, it seems that our natural characteristic of frankness was stamped upon the countenances of English boys. Gregory himself had a great desire to lead the mission, but he was destined to disappointment; for, having obtained the permission of the Pope, and his encouragement in the undertaking, he had secretly started from Rome, but before he had got far on his journey, with his band of Priests, messengers came furiously after them, saying that when the people of Rome found he was not in the city, they had assailed the Pope, and demanded his return, for no doubt the mob thought that the Pope had surreptitiously got rid of him, on account of Gregory being a great favourite with them. This was the reason that prevented Gregory from being the founder of our Church; but although foiled in his own immediate attempt, it did not hinder him from carrying out his main object, for as soon as he was raised to the see of Rome, he commenced looking about for some one who might undertake the mission he had been so suddenly stopped in. After some consideration, and after taking some time in determining on suitable men, he fixed on St. Augustine, Prior of the Monastery of St. Andrew, of which Gregory himself had been a monk, and sent him from the Eternal City with forty monks, to preach the Cross in Britain. There is still a picture in a chapel of that convent, of their departure. Before Augustine and his companions had gone many days journey, they got very frightened, by the strangeness of the way, and the rumours they heard of Britain and its inhabitants, so Augustine was persuaded to go back to Rome to pray that they might be excused. But Gregory was deaf to all his entreaties, for he said he had been prepared to face all the dangers that they would have to undergo; so they went on, and at last landed at Ebbs Fleet in the Isle of Thanet.

A farm house, on a piece of high ground rising out of Minster marshes, is still called Ebbs Fleet, and on a near approach, you can see at once, that the elevated land formerly must have been a promontory running out into the sea between the river Stour on the one side, and Pegwell Bay on the other. A field on the north side of the farm, is still shown as the spot where the landing of Augustine took place. To continue with our main subject: we have said that Augustine landed at Ebbs Fleet, he disembarked here that he might be safe on that side of the broad river, till he knew the mind of the king. There they landed "in the corner of the world," and waited there till they heard how the news of their arrival was received by Ethelbert, King of Kent. A review of the case of these men, set down as it were at the very ends of the world (to them as distant as New Zealand is to us) must be a great encouragement to our Missionaries starting off to some new and distant country to commence the great work, and the glorious consummation of Augustine's object must act as an incentive.

Here were Augustine and his companions, by name as Romans, amongst enemies, against whom they had so often fought, and they were also by creed Christians, so their case seemed a hopeless one.

We must now turn to Ethelbert. He was, so it is believed, great-grandson of Eric, son of Hengist, surnamed "The Ash," and father of the race of Kings of Kent, called "sons of the Ash Tree." His wife was the Princess Bertha, daughter of the king of Paris. She was a Christian, and on this the subsequent fate of England turned.

Hndson's Life of Queen Bertha says:—

"To undertake the task of tracing even a slight biographical sketch of Queen Bertha's life, would be to undertake an impossibility, for history does not furnish a clear outline. Nevertheless, it tells us quite enough to show that, as long as England is a nation, she should love and honour the memory of her first Christian queen.

"The silence deepens our respect, for all the biographical compositions, and the narratives of those times, are more or less imbued with absurd or revolting superstitions;

and yet it is from such sources that the historian must derive his material, when he begins his work at the commencement of modern history, on the confines of a prehistoric age.

"We should rejoice in the fact, that our Queen Bertha's name has wonderfully escaped all such legendary lore and fulsome flattery; that the title of Saint has never been appended to it. We should remember her name and her example, for she was the first who showed our Saxon forefathers what Christianity is; she was brought hither by Divine Providence "to make ready the way of the Lord." We may know much more about others who lived in later times, but let us not pass over Queen Bertha. As no connected memoir can be compiled, I have simply given an outline of the history, and drawn a picture of the period in which she lived. In the foreground of that picture I have placed Ethelbert, King of Kent, Queen Bertha, & Augustine, the Missionary Bishop who baptized Ethelbert. As a background to bring out the prominent objects and details of the picture, I have slightly sketched preceding history, beginning on the hazy indistinct horizon which separates the fabulous from the authentic.

"Having been led from the history of the Saxon Conquest into the history of the Church I have preferred giving the latter in the very words of authors fully competent to send forth information on so important a subject; and as I have extracted passages from Dr. Hook's 'Lives of the Archbishops of Canterbury,' I think it is well to give also a portion of his 'Introduction,' wherein he refers to the establishment of the See of Canterbury. He writes:— 'In tracing the present Church of England back to the Italian Mission and the See of Canterbury founded by

Ethelbert, and of which Augustine was the first Archbishop, we are not forgetful of the existence or the claims of the British Church or of the Celtic Christians, to whom, no less than to Augustine and his followers, we are indebted for the conversion of the Anglo-Saxon race. Who were the missionaries by whom the Celts, or first known inhabitants of these islands, were originally converted, as the event occured in a pre-historic period, it is impossible to decide.'"

No doubt Augustine thought Bertha might have some influence in persuading the King to allow him and his followers to settle and preach the gospel, and the sequel proves there was ground for this hope. Bede tells us that Queen Bertha had a chapel at Canterbury, "on the east of the city," on the site of which the venerable Church of St. Martin now stands. Bertha had brought from France a French bishop, Luidhard by name, and he officiated in this little building.

We quote from Queen Bertha and her times by Hudson, the following :—

"We can trace Luidhard's course almost as clearly as we can Bertha's, from the kingdom of Neustria or Soissons to that of Paris, and then on to Kent. And yet we have no positive evidence as regards the antecedent story of either the aged Bishop or the youthful Queen, till that day when they were fellow-travellers watching the white cliffs of France receding from their sight, and those of England growing into full proportion as they drew near to the unknown land before them.

"We may be sure that Queen Bertha was greeted by a hearty welcome; for a people who, even in their rudest primitive condition, were distinguished by the domestic virtues, could not be indifferent to the happiness of a monarch whom they loved and honoured. And Bertha's heart must have yearned towards her husband's heathen subjects; she must have remembered how Clovis, her own great ancestor, and France her native country, received Christianity through the medium of Clotilda."

"Thus we see that before Queen Bertha's times the ancient British Church had passed gloriously through the fiery trial of the Diocletian persecution; and it was then strongly contending against the paganism of the Saxons—retreating, but maintaining faith and courage.

"That was a time of tribulation, in the original meaning of the word; a threshing-time, when the wheat was separated from the chaff. Although the most ignorant portion of the Celtic population had clung to the soil, rather than to honour and religion, yet thousands of the nobler spirits had held fast the truth when they could hold nothing else.

"Abandoning their homes and their cherished places of worship, they had fled far away. The pastors had retreated with the flocks, and at last they had rested in strongholds fortified and entrenched by the Creator's skill, and they were building other churches, vestiges of which have been found among the mountains of Wales, or beneath the hills of drifted sand in Cornwall. Surely a special Providence has guarded those ruins, that they may, illustrate the history of the Early Church.

"The battered walls at Dover, which bore the form of the Redeemer's Cross when the land was polluted with idolatry, must have touched the heart of the Christian Queen of Kent; she must have grieved over the deserted house of God.

"Her thoughts about it must sometimes have been spoken to her children, and her words may have been 'cast like bread upon the waters' to be 'found after many days,' when her son Eadbald restored that church liberally endowed it, and built houses for the appointed Canons."

Very likely Ethelbert had been disposed favourably towards the new religion by Bertha, but his conduct on hearing that Augustine had arrived was hesitating; he would not permit them to come to Canterbury, and he would not under any conditions allow their first interview to be beneath the roof of a house, as he thought they might be wizards, and any charms they might exercise over him would, he thought, be less potent out of doors. The meeting of the two bodies was under an oak, and it must have been a very wonderful sight.

An ancient poet thus describes the procession:

> "Whiles Ethelbert was reigning King of Kent,
> St. Austin, sent by Gregory of Rome, Bishop,
> Landed in Tenet, with clerkes of his assent,
> And many monks to teach the faith I hope,
> That clothed were under a black cope,
> Which in procession with crosses and bells came,
> The Letanies synging in Jesus *his name*."

On the one hand, Ethelbert, "the Son of the Ash Tree," with his warriors, seated on the bare ground, and on the other, Augustine and his companions bearing a large silver cross, they, as they marched, chanting one of the Gregorian chants, which, even now are used, and which the pious Gregory himself composed. It is said that Augustine was a head and shoulders taller than any one else there. The king after hearing the message, which was explained by an interpreter, gave this very English answer, which shows how little we have changed in nature since that time; he said, "Your words and promises sound fair, but, because they are new and strange I cannot give my assent to them, nor can I leave all that I and my fathers and the whole English folk have believed so long. But I see that ye have come from afar to tell us what ye yourselves hold for truth, so we will not molest you, nay, rather, we are anxious to receive you hospitably and to give you all that ye want for your support, and ye may preach to my folk, and if any man will believe as ye believe, I will hinder him not."

"Such an answer simple as it was, really seems to contain the seeds of all that is excellent in the English character, exactly what a king should have said on such an occasion, exactly what, under the influence of Christianity, has grown up into all our best Institutions. There is the natural dislike to change which Englishmen still retain; there is the willingness, at the same time, to listen favourably to anything which comes recommended by the energy and self-devotion of those who urge it; there is, lastly, the spirit of moderation and toleration, and the desire to see fair play, which is one of our best gifts, and which, I hope, we shall never lose. We may, indeed, well be thankful, not only that we had an Augustine to convert us, but that we had an Ethelbert for our King." *Stanley's "Memorials of Canterbury."*

After this meeting the missionaries crossed over the broad ferry, from the Isle of Thanet to Richborough the "Burgh" or castle of "Rete" or "Retep;" here they were received formally by the King. Boys, in his Sandwich, says, "An old hermit lived amongst the Ruins in the time of Henry VIII., and pointed out to Leland what seems to have been a memorial of this, in a Chapel of Saint Augustine, of which some slight remains are still to be traced in the northern bank of the fortress." Here was also a head or bust, said to be of Queen Bertha, embedded in the wall, remaining till the time of Elizabeth; the curious crossing in the centre was then called by the common people St. Augustine's Cross. The good life of Augustine and his

Monks, their attention to the wants of the poor, and their devotedness to the Christian Faith, soon had its effect upon the King, who, we find, on the 2nd June, 597, was baptized, and it is in the highest degree probable that rite was performed in the old Church of St. Martin ; we find by the end of that year 10,000 Saxons had been baptized ; Fuller, in his Church History, says, "in the waters of the Swale." Ethelbert now soon retired to Reculver, and rebuilt the old Roman fortress there, giving up his own palace, and a Roman Church in its neighbourhood, to Augustine, to form a new Cathedral. These early missionaries soon associated their own names and those of their contemporaries with the Churches built; and being dedicated to a St. Andrew, St. Martin, St. Augustine, St. Lawrence, or St. Gregory, is of itself, in this neighbourhood, a sufficient evidence of great antiquity.

We cannot leave this interesting point in the history of our Christianity without remarking on the peculiar apathy displayed by the inhabitants of the Isle of Thanet and its vicinity on this subject, as if it were a matter of indifference whether Augustine landed here or anywhere else; and as though Julius Cæsar, having brought us within the pale of the civilised world, it signified nought that we could claim the spot he first put foot on. Shame on you, Thanet! how many lesser places have entered the arena of disputation on a point of much less significance, and with much less authority to help them to hold their ground!

We find in Hudson's Queen Bertha the following:

"All Canterbury pilgrims should visit Reculver; it is well to take that excursion on a bright autumnal day, for there is in that part of the country a good deal of woodland, which is seen to advantage when the leaves are changing colour.

And then the associations connected with what we have heard about departed centuries seems to harmonize with nature, when we feel that the present year is leaving us to join its predecessors of every age and period. It has dispensed the blessings with which it was charged, has filled our barns and storehouses, has even hung the hedges and bushes with red berries and feathery seed for the birds. Not one of God's creatures has been forgotten, and having done its work, the old year is peacefully passing away; bright are its latest leaves and lingering flowers.

"And as a year speeds on its appointed course, fulfilling the will of Him who ordained times and seasons, so it is with a century; with a cycle; with the whole span of time; with the span of each mortal life.

"Some such reflections as these may occur to those who go from Canterbury to see the ruins at Reculver on a fine morning in October. Losing sight of the woods, they come on an open level country, divided into large farms. The old-fashioned farmhouses, with their cheerful gardens full of chrysanthemums and China roses, look truly comfortable. Views of the sea are soon obtained, tranquil views; no busy port is near, but in the distance a vessel may be discerned, bound perhaps to some haven thousands of miles away.

"No one can fail to remark the tall twin towers which will guide him to the village on the cliff. Like the monks of old you may stable there, and take advantage of the hospitality afforded by 'the King Ethelbert' a most primitive little village inn. Its open door stands free to all, its windows peep through ivy leaves, and the swinging sign displays the name of that old Saxon King who held his Court at Reculver.

"Passing the inn, and some remarkably antiquated cottages, you come very abruptly to the edge of the cliff; for it has been falling and rolling under the lashing of the waves until the ruins and the signal towers, which used to be more than a mile from the edge, are now so near to it, that a sea wall is being raised to save them from destruction.

"The descent to the shore down that rugged slope is easily accomplished, and repays the trouble, for the cliff presents a pleasing picture, the colouring is so good. The earth of which it is composed is powerless to resist the strong towering waves, which often work here like battering-rams; but in that yellow soil vegetation flourishes; and all the green things are rampant there, in happy unconsciousness of the danger of their position. They seem to cling to that old cliff with the mural crown, as the young should cling to the aged, with a love which enlivens desolation, conceals weakness, and covers over the vacant places. As you stand on the shore, looking up to the broken wall surmounted by those tall sister towers, dark clouds should gather over the scene, leaving just sunshine enough to gleam upon the sea-gull's wing, and to soothe the old ocean into his 'chameleon-like' mood.

"Of course you will scramble up to see all that can be seen of King Ethelbert's palace. The northern wall of the massive fortress has been washed away, but the three other sides remain, overgrown by a gigantic ivy and a dwarf elder. Other trees grow among the ruins; some are stunted because their root-fibres are twining among Roman and Saxon remains.

"The church built by Ethelbert stood close to the castle, and they were together until the beginning of the 15th century, when the church was restored by the Lady Margaret St. Claire as a memorial of her sister, Lady Isabel.

"These orphan sisters, the nieces of an Abbot of St. Augustine's Monastery, were most happily united in their lives, supporting each other through many sorrows which the Wars of the Roses brought upon English families. They were on their way to visit a shrine of the Virgin at Broadstairs, when they were overtaken by a fearful tempest. The sisters were together in the storm, but one of them never recovered from the shock; and the survivor, seeking consolation in religious work, rebuilt the church on Ethelbert's foundations, and raised those sister towers to warn mariners from that dangerous coast. The church is again in ruins, but the signal towers are maintained by Government for the purposes of navigation.

"Looking round upon those hoary walls, all grey and mouldering with age, all tinged with red Roman brick, the casual observer cannot distinguish the era to which each fragment belongs, though it is easy to see that labourers of successive ages have there left signs of their handiwork, from the tile of Severus to the broad arrow of our Board of Ordnance.

"We look round with interest on the prospect which Queen Bertha looked upon, from what we imagine to have been her happiest home; that in which she lived after the conversion of her husband; and we search for the ancient stones placed there to build up that home; but the course of events, which marked the days she lived in, seems to tell us more about her than the old stones at Reculver can tell. Those events must have more or less shaped the circumstances in which she was placed, must have occupied her thoughts, awakening hopes and fears, causing joy and sorrow, thus affecting her daily life while she lived there as the wife and mother, as the Saxon lady and Queen; and those events were calculated to strengthen faith, to elevate hope, to enlarge charity and thus to penetrate the surface, and touch the inner life which Bertha led there as a Christian woman and heir of immortality. The impressions struck upon the hidden tablet of the soul may now seem indistinct and confused; but that kind of photography, beyond our human skill, is done by One whose eye cannot be dazzled, whose hand cannot tremble with a moment's hesitation. The features of individual character on each imperishable tablet will be brought out perfectly true to life, the very shades and colours that were given in time, fixed in eternity."

The Isle of Thanet is about nine miles in length from Sarre to Kingsgate, and in most parts five miles in breadth; it forms the north-east angle of the County of Kent; its situation is very bleak and open, especially towards the sea side, where there are few hedges or trees; the whole is in a very high state of cultivation, and has always been noted for its fruitfulness. This is to be attributed to various causes; first, the great amount of alga, or seaweed obtainable at only the expense of carriage, which, mixed with other dung, forms the finest manure, containing, as it does, so great an amount of vegetable and saline matter, which latter acts as a detergent to the land; but although good for the farmers, it is far from being pleasant, and he who comes here with sensitive olfactories must keep a sharp look out for these mixings when he takes his walks abroad in the summer time. Another reason for the abundance of their crops is the almost entire absence of hedges, which in some parts of the country destroy so much produce; here every inch of ground is made to yield to the utmost, and although the corn grows at the side of the road, we hear of very few complaints of its being knocked down, as in some districts would be the case, were this system to be tried. Again, the peculiar nature of the chalky sub-soil has much to do with the regularity of the crops, it acting in wet weather like a sponge, and in great drought supplying the land with moisture; thus, as a local couplet has it,

> "When England wrings
> The island sings."

for a very wonderful system of inhalation and exhalation is continually going on, and so complete does it appear to be, that the farmer in the Isle of Thanet can never have too great a quantity of rain, and neither does he ever complain of a dry season. It is very evident that the result is obtained by some of the above causes, and not by any virgin richness of the soil, as with the exception of a little land between Cliff's End and Sarre, it is not more than 9 or 10 inches to the chalk. An abstract from a Report of the Board of Agriculture, published some years since, will suffice to show this. "Much of it is naturally very thin light land, the greater part of it having belonged to the religious, who were the wealthiest and most intelligent people, and the best farmers of the time; no cost or pains were spared to improve the soil, the sea furnished an inexhaustible supply of manure, which was brought up by the tides to all the borders of the uplands, quite round the island, and most probably was liberally and judiciously applied by the monks and their tenants, and their successors to the present time have not neglected to profit by the example. Owing to these circumstances, Thanet was always, and most likely ever will be famous for its fertility. In short, there is not perhaps another district in Great Britain or the World, of the same extent, in such a perfect state of cultivation, or where land, naturally of so inferior a quality, is let for so much money, and produces such abundant crops. The tops of the ridges are about 60 feet above the level of the sea, and are covered with a dry loose chalky mould, from 4 to 6 inches deep, and is, without manure, a very poor soil. The vales between the ridges, and the flat lands on the hills, have depth of dry loamy soil from 1 to 2 feet thick, mixed with chalk, and of much better quality. The west end of the island, even on the hills, has a good mould from 1 to 2 feet deep, a little inclining to stiffness; but the deepest and best soil is that which lies on the south side running westward from Ramsgate to Monkton, it is there a deep rich sandy loam, and, being managed with care and expense, there is seldom occasion to fallow it. The soil of the marshes is stiff clay, mixed with a sea sand and small marine shells. There is no commonable land, nor an acre of waste land in the island." The old Kentish plough is persistently used by farmers in the Isle of Thanet and its neighbourhood, and certainly, if we may judge from their crops and land, they have no need to try another; but after seeing in the Midland Counties the newest invention in ploughs, cutting up the ground like a knife, and only stopping now and then for the driver to have a suck at his wooden beer barrel; it seems exceedingly strange to see the ploughman stop at the end of every furrow to take his coulter out and reverse it, which of itself hinders apparently half as much time as the ploughing takes. The best explanation given by practical farmers is that the old plough breaks the earth as it turns over, and that, in consequence of the land being light, it suits it better than having it packed in slices one against the other as the iron plough would do. Another peculiarity is, that, after a clover or sanfoin lay, the ground is ploughed up, manure is thrown upon the ploughed land and harrowed in; this is certainly peculiar to this part, but all farmers here persist in its use; and the effect is not such as to induce them to discontinue it.

Fishing forms the principal occupation in winter of the lower class in Ramsgate, Margate, and Broadstairs, many of the fishermen earning during that time, enough to assist them through the summer months; many also during the season, make a harvest with their pleasure boats, which at that time are in constant requisition. It is much to their credit that, in cases of distress, they are ever willing to hazard their own lives to save others—it cannot be said for gain—the reward in most instances being only the small amount given to each man every time the life-boat goes out; but we seldom hear a complaint that more might have been done, although there are plenty of dissatisfied people to find fault if it can be found, vide the case of the Broadstairs boatmen, on the terrific night, in Oct. 1859, *(the "Royal Charter" night,)* when, after several ineffectual attempts, at Kingsgate, to rescue the crews of two vessels wrecked there, it was said they were intoxicated and did not try; but this was proved by the Report of a Commission instituted to enquire into the truth of the case, to be without foundation. The fact of a public looking on (if not at the time, in the report of the case at least,) and that public, one that makes little allowance for failures, no doubt stimulates the men to increased exertion. But there is also another thing, a very appreciable rivalry between two neighbouring boats, as to which shall be the first to reach a wreck; no doubt this is partially to be attributed to the salvage claimed by the first boat to reach it, but it is nevertheless an ascertained fact that this rivalry has been the means of saving many lives. A case, illustrating the view that it is not entirely for gain that these men hazard their lives, occured in February, 1860, when the Spanish brig, *Samaritano*, got on shore at the Wedge, near Margate, the life-boat men having gone out from that place to her rescue, the Ramsgate men were apprised of the danger of the Margate crew, they therefore started off in one of the worst days of that winter to assist them, notwithstanding the blinding snow squall, although they well knew the reward would not be theirs; and after being out in that small boat, snow falling and freezing as it fell, continually drenched, in fact, hardly being otherwise than covered with water for seven hours, their task rendered doubly difficult through not being able to approach the wreck to leeward, they returned, having saved both crews. A very brave rescue occurred in January of the year 1881. A large vessel, named the *Indian Chief*, having struck on the Long Sand, a dangerous place about 30 miles north from the coast of the Isle of Thanet, the Ramsgate lifeboat being telegraphed for by the Harwich harbour authorities, was tugged out about twelve a.m., and not being able to find the wreck, the men were all night in the teeth of a storm of wind, which blew the sea water all over them, and as it fell it froze on their faces; when morning at last came, they saw the ill fated vessel, and after facing very great hardships, they at last saved eleven out of a crew of twenty-eight men. Surely such gallant conduct as this should disarm the faultfinders, and be amply rewarded, which, unless by a Subscription, it seldom is now.

The Volunteer, and nearly all are now such, although we are a nation of shopkeepers, will find, both at Ramsgate and Margate, excellent ranges for practice, although rather exposed, but being at the seaside and shooting across the sea, it is hardly to be expected otherwise. In connection with the Ramsgate Corps a Club has been formed for practice; this Club meets fortnightly throughout the year, and all members or honorary members of any corps are eligible to be admitted. The Range of the Ramsgate Corps is situated at Cliff's End, beyond Pegwell.

Originally the Island was partially covered with wood, hence the names of the Villes Northwood, Southwood, &c.; but at the present time there is very little wood, excepting near Minster where the soil appears a little deeper. These woods were of great benefit to the inhabitants of this Isle, for it was to them that they used to retire with their wives and families upon the approach of the Danish Pirates.

Some years back a great many flint implements were found in the drift here, some shaped as arrows and darts, others in the form of a chisel.

Mr. Seddon in his Introduction to Ancient Examples of Domestic Architecture in Thanet says:—

"Many years ago my attention was called to the interesting remains of old domestic architecture in the Isle of Thanet, and particularly to those in the quaint village of Reading Street, by Professor Donaldson, who made some geometrical elevations to scale of several of them. Nevertheless, when more recently professional engagements took me into their neighbourhood, their remarkable character burst npon me as a surprise, and made me wish that they could be presented to the public in a form which could give some fair idea of their modest charms and peculiarities."

To those who are interested in this subject we would advise getting a sight of the above book which is now very scarce, but can be seen at *Wilson's Library.*

The Life-Boat

ON THE COAST OF KENT.

We will begin by stating that the first person who wrote on Life Boats was Lionel Lukin, a coach builder of London, in 1785, who introduced a very primitive one in that year; this was the first used, and after its launching at Bamborough, was the means of saving many lives. Lukin, like many another inventor, was doomed to disappointment, as he could not get the Admiralty's ear, and so he died not only unrewarded but actually believing that his great invention would never be brought into active use; but in 1789, the Public were stirred to the heart by such an occurrence, as the wreck of the *Adventure*, Newcastle, where thousands stood looking on at the terrible sight and could do nothing to save the poor creatures, that premiums were offered, and eventually Greathead was selected to build for Government some improved Life Boats. At this time, 1803, the Duke of Northumberland took a great interest in the subject, and in 1822 Sir William Hillary came forward as the Champion of the Life Boat, and he, in conjunction with Mr. Thomas Wilson and Mr. George Hibbul, was the founder of the Royal National Institution for the Preservation of Life from Shipwreck, which took its birth from 4th March, 1814, the objects of which are:—

"*1st*, To station lifeboats, fully equipped, with all necessary gear and means of security to those who man them, and with transporting carriages on which they can be drawn by land to the neighbourhood of distant wrecks, and to erect suitable houses in which the same may be kept.

"*2nd*, To appoint paid coxswains, who have charge of, and are held responsible for, the good order and efficiency of the boats, and a quarterly exercise of the crew of each boat.

"*3rd*, To liberally remunerate all those who risk their lives in going to the aid of wrecked persons, whether in lifeboats or otherwise; and by rewarding with the gold or silver medal of the Institution such persons as encounter great personal risk in the saving of life.

"*4th*, By the superintendence of an honorary committee of residents in each locality, who, on their part, undertake to collect what amount they are able, of donations towards the cost, and of annual contributions towards the permanent expenses of their several establishments."

The following taken from " The Lifeboat and its Work " is a specimen of the labours of Boatmen in these parts.

MARGATE.—About half-past three o'clock on the morning of the 25th of Jan., 1871, while the wind was blowing strongly from the east, and during a heavy snow-storm, the brig *Sarah*, of Sunderland, bound from that port to Southampton with coal, and having a crew of six men on board went on the Margate Sands. The wreck was not observed from the shore until about noon, the hull of the vessel being under water. As soon, however, as it was noticed, the *Quiver* Life-boat was immediately launched, and proceeded to the spot, when the crew were found to have taken refuge in the foretop. With some difficulty, on account of the heavy sea running alongside the wreck, the six men were happily rescued from a watery grave; two of them were very severely frostbitten in the legs, and it was not without much difficulty and danger that they were got into the Life-boat. However, the efforts of the boatmen ultimately proved successful, and all were safely brought ashore, and the two injured men at once placed under medical treatment. The snow was lying some inches deep in Margate at the time. A lugger had attempted to get to the rescue of the shipwrecked crew, but was unable to get sufficiently near to the wreck, through the heavy breaking seas, to render assistance; and, no doubt, the poor men would have perished in the absence of the Life-boat and her gallant crew.

About a fortnight later, the same Life-boat was able to effect the rescue of another shipwrecked crew. A strong gale from the E.N.E., was experienced there on the 10th of February, and, about ten o'clock at night, a vessel was observed burning a tar barrel as a signal of distress, she apparently being on the Walpole Rock. The *Quiver* was at once taken along the shore on her transporting-carriage to the lee side of the Longnose Rock, and launched through a very heavy surf to the vessel, over which the sea was then breaking. With some difficulty the crew of nine men were taken off by the Life-boat. The vessel was the brig *Thessalia*, of Whitby. The horses used to draw the Life-boat on its carriage, although accustomed to the work, could hardly be got to take the boat to the water's edge, on account of the strength of the wind and heavy rollers setting in, which at times completely covered them.

Ramsgate and Broadstairs.—At daybreak, on the 28th of March, 1871, during a north-easterly wind and a heavy sea, a large barque was seen ashore on the Goodwin Sands with a signal of distress flying. The Life-boat *Bradford* and the harbour steam-tug *Vulcan* were at once despatched to the spot, and on arriving there found the Broadstairs Life-boat, the *Samuel Morrison Collins*, had also just arrived. Both boats then went alongside the barque, which proved to be the *Idun*, of Bergen, bound from Newcastle to Venice with coals. The crew of fourteen men, together with the son and daughter of the master, were then taken into the Life-boats; but on returning to shore the boats unfortunately grounded on the sands, where they had to remain, in the midst of great danger, three hours, until the flood tide made, when they were taken in tow by the steamer, and arrived safely in Ramsgate Harbour about three o'clock that afternoon.

The master of the vessel afterwards expressed, through the columns of the *Times*, his acknowledgments for the valuable services thus rendered to himself and the others on board the wrecked vessel. His letter was as follows:—

WRECK ON THE GOODWINS.

To the Editor of the Times.

SIR.—Will you kindly allow me, through your widely circulated journal, on behalf of myself and crew (sixteen in all) of the bark *Idun*, 670 tons, of Bergen, Captain Meidell, from Newcastle, bound for Venice, with a cargo of coal, wrecked on the Goodwin Sands on the night of the 27th March, in a strong gale from the northward and eastward, to express my heartfelt thanks and deep sense of gratitude to the brave and gallant crews of the Ramsgate and Broadstairs Life-boats and the Ramsgate steam-tug *Vulcan*, for the invaluable services rendered to us under circumstances of very great distress and danger on the Goodwin Sands, which resulted in the preservation of all our lives?

The noble boats, under the able and skilful management of their persevering crews, came out from the land at daylight in the morning, and dashed fearlessly into the foaming breakers, crossed the boiling sand, and at very great risk to their own lives (the sea breaking heavily into the boats as they approached) succeeded in reaching the ship and laying alongside to our rescue. We were all then hastily, but most kindly, assisted into the two Life-boats (my daughter being with me, a passenger) together with a quantity of nautical instruments, clothes, and other effects. At this time the tug was waiting to windward at the edge of the sand, near the breakers, to receive the boats: but the boats, on leaving the ship, could not get off the sand, the tide not having flowed sufficiently to enable them to pass through the breakers; they were, therefore, compelled to wait and allow the boats to beat over the sand to leeward through the boiling sea, breaking heavily into them for three hours, when at length they succeeded in getting off the sand, where the steam-tug (having come round to leeward) was waiting in readiness to receive them.

The tug then took the two Life-boats in tow (one being disabled in her rudder) and steered for Ramsgate Harbour with flags flying from their mastheads, where we arrived at 3 p.m. We were received with loud shouts of joy from hundreds of English spectators on the pier, who had assembled to welcome our safety to land. We were then conducted to the Sailors' Home, where refreshments were already prepared in readiness for us, which we found most welcome after an exposure of about fourteen hours to wet and cold, and where we still remain for the present. We left the ship full of water with her mainmast gone, and no prospect of her ever coming off the sand.

Permit me, sir, in conclusion to say that too much praise cannot be given to the English nation for the introduction and success which has attended their noble service, THE LIFE-BOAT INSTITUTION, established for the preservation of shipwrecked mariners of all nations.

Your insertion of these my grateful acknowledgments for the invaluable services rendered us will greatly oblige

Your obedient Servant,

H. C. MEIDELL,

(Captain of the said bark, Idun.)

Again, on the 19th March, 1872, the Ramsgate Life-boat and Steamer, and the Broadstairs Life-boat *Samuel Morrison Collins* saved the crew of eight men from the brig *Defender*, of Sunderland, which was wrecked on the Goodwin Sands during a fresh gale from the N.E., and in a very heavy sea. The Broadstairs Life-boat was the first to arrive at the scene of the wreck, but in running towards the vessel she was struck on the broadside by a tremendous sea, which in a second threw her on her beam-ends to the momentary consternation of her crew. She however behaved nobly, and elicited from her brave fellows their hearty approval. She was full of water, but that was soon discharged. By this time she had been driven to leeward of the wreck, and the anchor had to be dropped, prior to a fresh attempt being made to go alongside, meantime the Ramsgate Life-boat had been towed to windward, both tug and boat shipping much water, and the paddlebox of the steamer being damaged by the heavy seas. The Life-boat was then slipped, sail was made, and when she neared the vessel the anchor was let go, and she veered down abreast of her, when a line being thrown on board, she was hauled alongside between the heavy seas that were breaking over the wreck. In doing this, much risk was incurred; the boat, indeed, having her stern damaged through being dashed against the side of the brig. With much difficulty the master and seven of the crew were saved by the Ramsgate Life-boat; there was one man left, he got into the ship's boat, and slipped down to the Broadstairs Life-boat and was saved.

WALMER, KENT.—Most admirable service was performed on the morning of the 16th Oct., 1872, by the Walmer Life-boat *Centurion*, as will be seen by the following report furnished by her coxswain :—He says, "It was blowing hard from the southward, with a heavy sea on the beach when I observed a vessel on the Goodwin Sands; I immediately assembled the boat's crew and launched the Life-boat, and proceeded towards the Sands under a reefed storm foresail. On crossing the South Sand Head in a tremendous sea, the boat filled seven or eight times, and two of our men were nearly washed overboard. After crossing the Sands we kept away towards the wreck, and on nearing her, saw she was full of water, with the sea making a clean breach over her. Feeling it would be dangerous to go alongside, we let go the anchor to windward and dropped down towards her; we could see the crew huddled together before the foremast, with the seas breaking over them. On reaching as near as possible, we managed—with the assistance of the loaded cane and line—to get a rope to the vessel, and each man fastening it round his body, we hauled them through the broken sea; but the foremast going, and the seas running higher, when two only had been saved by this means, the remaining two men took to the mainmast, where there was great difficulty in communicating with them; but in about half-an-hour the maintopmast rigging gave way, and having hooked the wreck of this, the men were induced to slip down it into the sea, and get hold of the rope that we had secured to the wreck. In this way the remaining two men were saved, making in all four men, the entire crew of the vessel. They were very much exhausted when taken into the boat. The wrecked vessel was the schooner *Hero*, of London, bound from Newcastle to Truro, with a cargo of coals. In less than five minutes after the men were rescued from their perilous position, the wreck disappeared, and there was not a vestige of her to be seen. We lifted our anchor and proceeded towards the shore, where we hove up at 11 A.M., in the presence of a concourse of people who took hold of the capstan rope and hauled the boat up to the boat-house amidst the cheers of the people." The crew spoke in great praise of the performance of the Life-boat on this occasion.

KINGSDOWNE, KENT.—This Life-boat performed, in conjunction with the Walmer Life-boat, a very gallant and praiseworthy service a year or two since, as will be seen by the following account, which is taken from the depositions made by the coxswains and crews of the Walmer and Kingsdowne Life-boats :—

"On the morning of the 17th Dec., 1872, we were summoned by the firing of minute guns and other signals of distress from some vessel on the Goodwin Sands, and at 3 A.M., we launched from Walmer and Kingsdowne simultaneously in the *Centurion* and *Sabrina* Life-boats, the wind blowing heavy from S.S.W., weather thick with rain. We proceeded in the direction of the signals, and, after encountering a fearful sea, we discovered a large steamship ashore on the inner part of the Goodwin, known as the Callipers. At 4 A.M., boarded the said vessel, which proved to be the *Sorrento*, screw steamship, from the Mediterranean, with a cargo of barley, and bound to Lynn. The master asked us to remain and float the vessel if possible. We put on board the greater part of both Life-boat crews, who threw over cargo and carried out an anchor, with a view, if possible, of floating her off the Sands at flowing tide, but the wind and sea increasing, as the tide flowed she soon became a total wreck, filling with water, and the heavy broken waves making a clean breach over her. At

11 A.M., thinking the two Life-boats, the *Centurion* and *Sabrina*, were insufficient to rescue the whole of the steamer's crew, her ensign was hoisted, Union down, for more assistance, but none came, and at noon the *Centurion* Life-boat, which was then lying alongside, together with some of the steamer's boats were swept away, and the Life-boat was much damaged in her bows by a huge wave breaking bodily over the steamer, sweeping all before it, and causing some of the ship's boats to come into collission with the *Centurion*, which was immediately swept, with the rest of the floating wreckage into the surf, and to the back of the Sands altogether, leaving the greater part of their crew on board the steamer. The *Sabrina* Life-boat was anchored a short distance to windward, and the coxswain seeing the disaster happening to the *Centurion*, and feeling assured that a heavy loss of life must immediately follow, and that amongst the sufferers must have been his three sons, who had voluntarily accompanied him in the Life-boat, and were put on board the steamer, to try and float her from off the Sands, ordered the *Sabrina* to be immediately run alongside, though it should cost his own life and the rest of his boat's crew. This act was so successfully performed that the steamer's captain and his crew of twenty men, together with the pilot and the Life-boatmen, immediately leaped on board the *Sabrina*, which, with the whole party of no less than forty-six persons, immediately sheered off, set a close-reefed foresail, and steered through the heavy boiling surf to the off edge of the Goodwin, where our brethren in the *Centurion* were awaiting us at anchor, and to whom we transferred a necessary portion of the steamer's crew and Life-boatmen from the *Sabrina*, and then immediately proceeded, in company, across the

Sands in a very heavy sea, round the North Sand Head for Broadstairs, where we arrived in safety at 2-15 P.M."

The REV. J. GILMORE, late Rector of Trinity Church Ramsgate, also in his "Storm Warriors" gives some very graphic pictures of life-saving by the crews of our Lifeboats, to which we would draw our reader's attention; his original chapters were written for Macmillan's Magazine, and were also published in the National Life-boat Institution's papers, Mr. Gilmore says in his preface. Mr. Macmillan, the publisher, writes: "The Public have evinced considerable interest in those tales of Life-boat work," and so invited, Mr. Gilmore set about to collect matter for this his great effort, not by risking his life in the Life-boat, but he says, "I have managed better, I have had sometimes two, three, or four boatmen up to my house, and we have fought their battles over again; questioning, and cross questioning, getting particulars from them small as well as great."

"What did you do next?" To one such question, I remember the answer was—"Why then we handed the jar of rum round, for we were almost beaten to death."—"But with the seas running over the boat, and the boat full of water, it must have been salt water grog very soon —how did you manage it?"—"Well, Sir, when there was a lull, a man just took a nip; then if there was a cry, 'Look out! a sea!' he put the jar down between his legs, shoved his thumb in the hole, held on to the thwart with his other arm, then bent well over the jar and let the sea break on his back."

Thus getting them to recall incident after incident, I got the full details of each adventure; and when we arrived at the more stirring scenes, it was very exciting work indeed; the men could scarcely sit in their chairs—their muscles worked, faces flushed, and most graphically they told their tales, I, not one whit less excited taking notes as rapidly as possible.

Truly I must live to be an old man before I forget the hours I have spent in my study with Jarman, Hogben, and Reading, and R. Goldsmith, and Bill Penny, and Gorham, and Solly, and some other of my brave boatmen friends, as they have told me their many experiences and toils and dangers in life-boat work.

Take, as an instance, his account from the same book. "The harbour steam-tug *Aid* and the life-boat had started from Ramsgate early in the day, to try and get to the *Northern Belle*, a fine American barque, which was ashore not far from Kingsgate; but the force of the gale and tide was so tremendous, that they could not make way against it, and were driven back to Ramsgate—there to wait until the tide turned or the wind moderated.

"About two in the morning, while they were making ready for another attempt to reach the *Northern Belle*, rockets were fired from one of the Goodwin light-vessels, showing that some vessel was in distress on the Sands. They hastened at once to afford assistance, and got to the edge of the Sands shortly after three in the morning. Up and down they cruised, but could see no signs of any vessel.

"They waited until it was daylight, and then saw the upper portion of the lower mast of a steamer standing out of water. They made towards it, but found no one was left, and no signs of any wreck floating about to which a human being could cling."

The undermentioned Books are recommended as giving full detail and account of 'The Life-boat and its Work,' both on this and other of the rock-bound coasts of England.

BALLANTYNE'S LIFE BOAT.

BALLANTYNE'S LIGHT HOUSE.

BALLANTYNE'S FLOATING LIGHT OF THE GOODWIN SANDS.

LEWIS' LIFE BOAT AND ITS WORK.

GILMORE'S STORM WARRIORS.

In these and other kindred works, adventures equalling any romance will be found, and visitors to the sea-side will well spend their time in perusing the same, when probably they will be induced on their return to country quarters to send their mite to the NATIONAL LIFE-BOAT INSTITUTION, or the SEAMENS' or SMACK BOYS' HOME at Ramsgate

Ramsgate.

RAMSGATE can boast few historical associations, albeit the inhabitants at the time of its rising popularity endeavoured to prove that its name Ramisgate (Ruim's Gate) or Romsgate was a corruption of Roman's Gate, it is a question whether the Sea Gate or way through the cliff was formed at the time of the Romans' visits to this Island. From Furley's "Weald of Kent" we find records as far back as the 13th century, and Canon Robertson speaks of many who took their names from the place, as Martin de Ramisgate, Clement de Ramisgate, &c. From a state paper of Elizabeth's reign it has also been discovered that the Hope Cliff's End was at that time a great trouble to mariners, many wrecks being recorded as taking place near the entrance to the Stour. Others take it as derived from the old Norse name—*ram*, strong; and *gaeta*, gate; and suggest that this strong gate was used by prowling pirates anterior to the time of the Romans. But if it was, it must have been a very different thing to the Gate of the present day, as Boys in his Sandwich speaks of it as leading between Hereson and Ramsgate, and being only six feet broad. The fact of the ville of Ramsgate, in Lewis's time, being composed of Northwood, Southwood, Court Stairs leading to Pegwell, and Hereson, seems a plain proof that the Town has increased seaward of late years, which increase would probably begin at the time of the gate being formed; an old writer says, "according to the return made by Archbishop Parker, to the privy council in the year 1563, it appears that there were then 98 households in the ville, but owing to the prosperity of Ramsgate it has greatly increased, insomuch that in 1773, there were in St. Lawrence, including Ramsgate, which contains more than two-thirds of the houses and inhabitants of the whole parish, 699 houses, and 2726 inhabitants; and in 1792, there were 825 houses, and 3601 inhabitants; which is a great increase for so short a space of time." Undoubtedly its convenient situation and close proximity to the Downs induced the Parliament, early in 1749, to commence a Harbour here for the shelter of ships, not exceeding 300 tons burthen; the work was carried on with spirit for some years, when some disagreement as to the plan occasioned a stoppage, and it was not until 1761 that it was recommenced, when it was completed according to the original design.

In 1788, the advanced East Pier was commenced, to shelter the harbour during gales of wind, which certainly had a beneficial effect, for, in 1795, 300 sail took refuge therein, some being vessels of 500 tons. Between the years 1792 and 1802, a Lighthouse at the end of the West Pier, (which has since been rebuilt) the Watchhouse on the East Pier, and the Harbour Master's House, and adjoining building, now used as a Custom House, were erected. The expense of constructing this Harbour, &c., amounted to £700,000, but the amount was well expended, when we see how many valuable lives are saved, by vessels taking refuge here, annually. The area of the Harbour is about 46 acres; the piers, chiefly constructed of Purbeck stone, are 26 feet in breadth, including the parapet fronting the sea; the length of the East is 2000 feet, that of the West

1500. The Slip Way, for repairing vessels, is a source of much amusement to those who visit here in the autumn or winter, when many vessels come in requiring re-coppering or other repair. The arrival of the Fishing Smacks is also witnessed by many early risers, who often pick up a string of cheap fish; other later ones take a boat and go out fishing, but this they will find is not the way to eat a cheap fry, but a very pleasant day's amusement. During the herring season, in the autumn, the excitement is very great, the east gully being full of boats awaiting their turn to unload. Some idea may be gathered of the amount of traffic from the fact that, during the height of the fishing season, in one day 150 tons of herrings will be despatched by rail to London, and by other conveyances to Sandwich, Canterbury, &c. The principal fish caught are turbot, brill, soles, plaice, cod, whiting, &c. It is no uncommon thing for a boat during the herring or mackerel season, to make £60 or £70 for one night's work; on the other hand, they are sometimes a whole week without taking anything, not being able to meet the shoal, and of late years the mackerel takes in spring have been but small. The yachts used in summer time for pleasure excursions are also rigged up in winter for fishing and hovelling, which is looking out for wrecks and lying to, with the hope of giving help, and so getting salvage. Population in 1881, 16,234. St. Lawrence, now part of Ramsgate, 6,449. Ramsgate is 72 miles from London, 16 from Canterbury, 30 from Ashford, and 20 from Dover.

ENTRANCE TO TOWN. Issuing from the South-Eastern Railway the visitor passes down Chatham-street, where are Townley House Seminary, conducted by the Misses Kennett, the Proprietary College, and Chatham House, by Rev.

E. G. Banks. This School is now one of the principal features of attraction in the Town, having been re-built at an enormous expense by the present proprietor, as well as some very ornate houses adjacent; as a private School, this now perhaps stands unequalled, every modern convenience being adopted to ensure comfort as well as health; the grounds are very extensive; the cricket displayed by the members of this School ensuring the best teams in the county as opponents, and on their own ground the scholars are nearly invincible. The re-building of this School has allowed of the widening of Chatham-street, which, from its being the only access from the South-Eastern Railway and also from Margate, was occasionally in the season completely blocked, now it is one of the boldest approaches, only requiring the same improvement continuing to the Market Place. Turning into High-street we pass on by the Parish Church of S. George, built in the florid Gothic style, surmounted with an hexagonal lantern tower; on the right, the Congregational Chapel, S. George's Hall, the new pile of buildings, Theatre, etc., erected by Mr. Sanger on the old Skating Rink site, and Post Office; then just out of the High-street, approached by way of Turner-street, is S. James's Theatre, a perfect bijou theatre; further down we come to the Town Hall, thence by Harbour-street to the Sands or Pier.

Do not leave the Pier without a visit to the Camera Obscura, which may be seen at any time during the day. It is needless here to explain what it is, as most visitors will understand the construction, but from its peculiar situation at the Watch-house on the West Pier, commanding a view of the whole front of the Town, together with the Harbour, Pier, Basin, etc., as also the Downs and Deal, it is a source of endless amusement, as, with all the truth of a photograph, it gives the colours of Nature, the movement of everything within its focus, let it be ever so minute, and the distinctness of a painting. A shilling is often paid for viewing a Picture or Panorama far from being natural, here you have a Panorama painted by nature itself, with movable figures, and at a lower price. The painter will here be able to see and study the rippling of the water on the canvas, as it is just what so many in vain attempt to depict; when the sea is rough and vessels are making for the Harbour, the effect is hardly to be described, but must be seen to be admired. In the morning there are the London boats to see off, in the afternoon and evening the same to see arrive, these form, with the trips in Waggonettes round the Isle of Thanet, to Margate, Pegwell Bay, and the Saltpans for Richborough Castle, the principal sources of amusement. The visitor may now take a walk to the end of the Pier, and see the whole of Ramsgate, or at least that part known by those who come during the season; looking towards the Town, on the extreme right may be seen the Coast Guard Station, Granville Terrace, the handsome Granville Hotel, Gardens, and Marine Mansions adjacent, with the Marina at the base of the Cliff, new Promenade Pier, Victoria Parade, the Augusta Stairs, Wellington Crescent, Albion Place, the London Chatham and Dover Railway Station, and the Colonnade; on the left the Roman Catholic Church of S. Augustine, the Monastery, Royal Crescent, Christ Church, the Paragon and Paragon Baths, Nelson Crescent, Prospect Terrace, Sion Hill, &c. Standing at the Pier Head the panoramic effect at night during the season is remarkable, every window being lit up, and the long row of lights from extreme west to east culminating in those along the

Victoria Gardens and down the Marina Road to the Sands and Station, which give the appearance of a Parisian garden on a Féte night. But a very different state of things is to be seen here during the tempestuous weather experienced in winter months, the pier extending so far into the sea that vessels endeavouring to enter are completely covered with spray, and so extremely difficult is it, from the tide at times setting strongly round, that many vessels are carried on to the pier head, and make a job for the shipwrights; others not making sufficient allowance for tide, are taken round the east pier, and driven at the back of it, where they are almost certain to go to pieces, but luckily the visitor is not often called upon to view such calamities as the latter; in 1860 two went round together.

On the East Cliff are steps called Augusta Stairs, leading from the East Cliff down to the Sands, (the humours of which have been so well pourtrayed by W. P. Frith in his picture of "Life at the Seaside,") the Station of the London, Chatham, and Dover Railway, which line has brought us nearly 20 miles closer to London than formerly, given a good approach to the Sands, a massive Sea Wall and Esplanade, which gets rid of the loose sand and rubbish that originally was so great a nuisance between the breakwaters, and which is now firm quite up to the East Pier wall; on the cliff, facing the Sea, are Albion Place, Wellington Crescent, Granville Hotel and Victoria Gardens. This Hotel, designed by E. Welby Pugin, is well worth the attention of all visitors whether they seek within its doors the comforts of a first-class hotel, or the delights of a Turkish Bath; these, together with sulphur, ozone, and other baths, may be had here in perfection. The originator has had a worthy follower, for no expense has been spared in the elaboration of plans, while the Marina and Etablissement, which was only a conception of the former, has proved in the hands of the latter one of the most attractive pleasure resorts in this fashionable watering place; and we can only wish that this Hotel may be as well filled all the year round as it is now during the summer months. Beyond this is the Coast Guard Station, New Promenade Pier, a very light and handsome structure, and East Cliff House, (for many years the residence of that good and philanthropic gentleman, Sir Moses Montefiore,) the grounds of which, to the extent of 13 acres, extend quite down to the cliff. At the rear are the Jewish Synagogue, and the Mausoleum erected by Sir Moses at the decease of Lady Judith Montefiore, his wife; also Jewish Almshouses erected by his munificence. Midway between here and the Margate Road is the Cemetery, the South Eastern College; and, in the Margate Road, the Isle of Thanet Steam Flour Mill.

The Visitor, starting from the gate of the Harbour and moving westward, will pass the *Castle* and *Royal Oak* Hotels, the National Provincial Bank, the *Royal* and the *Royal Albion* Hotels, Hammond & Co's. Shipping Office, (Lloyd's Agent has also Rooms at the same office), the New West Cliff Arcade and block of Mansions. Then, ascending the Cliff to Sion Hill, a remarkable bird's-eye view is obtained of the Royal Harbour, the Downs, Sandwich, and Deal Pier in the distance, with, on a very bright day, the Cliffs of La Belle France looming in the distance; then, passing along Nelson Crescent, the Paragon, and Royal Crescent, the Church of S. Augustine comes in view, the masterpiece of the great Gothic revivalist, A. W. Pugin, and the Grange, late the residence of the Pugin family, also built by the same architect; then, close at hand, only separated by the roadway, is the Monastery, built by the munificence of the late Rev. A. Luck, and designed by E. W. Pugin, the architect of the Granville.

No visitor who can stand a sail but takes a trip to those (by all mariners) much dreaded Goodwin Sands, which, about 7 miles distant, extend from a little to the eastward of the Harbour, directly in front of it, in the direction of Deal. At night you will easily distinguish them by the four Light vessels: the Goodwin Light, the Gull Stream, the South Sand Head, and the East Goodwin. These Sands are visited daily by an immense number of people from Ramsgate and Broadstairs, trips being made by yachts and sailing boats almost daily; sometimes at low tide the boat will land its passengers, who will traverse many miles of these arid sands, and occasionally play a game of cricket there Then there are trips to Boulogne and Calais once or twice a week; also trips by rail to Dover, Canterbury, Deal, Walmer, Hastings, &c., daily. The subjoined account of the Goodwin Sands is from Hasted's Kent. "Opposite to Ramsgate, at the distance of somewhat more than two leagues from the shore, are the Goodwin Sands, which extend in length from north to south almost ten miles, and in breadth about two, and are visible at low water. Though these Sands form a bank, which in conjunction with the North and South Forelands, renders the Downs a tolerably safe harbour, yet in general they are very destructive to navigation; ships striking on them seldom escape, being usually quite swallowed up in a few tides, and sometimes in a few hours."

At the entrance to the East Pier stands the Obelisk, erected in commemoration of His Majesty George IV's.

embarkation for Hanover and safe return. It was erected by private subscription; the proportions are two-thirds the size of the largest of the celebrated two at the entrance of ancient Thebes.

Ramsgate, with its added area, which in October, 1878, an Improvement Bill was passed to legalize, is now a Corporate Town, having a Mayor, Aldermen, and Councillors, and is severed from Sandwich, a limb of which it had been for so many generations.

THE TOWN HALL, which is situated over the Market-place, is supported on columns, and was erected in 1839. It will be apparent to any casual visitor that this antiquated structure should be pulled down and rebuilt, so as to give more room, as at present it is not an infrequent occurrence for the Hall to be wanted for the Corporation, County Court, Magistrates, and Coroner at one time. It contains an original portrait of the Queen by Fowler, placed there in 1840. The likeness is one of the best ever executed of Her Majesty—the character dignified, and the colour brilliant without gaudiness. It also contains an excellent portrait by the same talented artist, of the late R. Tomson, Esq., for many years the high constable of this town; another of the late deputy, T. Whitehead, Esq., and his successor J. Webster, Esq., both by Tweedie; also one of Sir Moses Montefiore, by Captain Weigall.

THE VEGETABLE AND FISH MARKETS are situated under the Town Hall.

S. GEORGE'S CHURCH.—Though the Town of Ramsgate has been from ancient times a distinct "Vill" within the liberties of the Cinque Ports, a member of the Port of Sandwich, and charged with the maintenance of its own poor, yet till 1826 it formed ecclesiastically a part of the extensive parish of S. Lawrence, and had but one Chapel of Ease (now S. Mary's), consecrated in 1791. In 1826 a special act of Parliament was passed (in conformity with the general Act of George III. 58, providing for the building of additional churches in populous parishes), whereby not only was the Church of S. George's built, but the old Vill of Ramsgate was "in consequence of its increase in extent of buildings and number of inhabitants separated for ever from S. Lawrence, and made a distinct Parish, and Parish Church." By the revenge of time Ramsgate, which once was an ecclesiastical part of S. Lawrence, has now in its civil aspect swallowed up S. Lawrence, which forms two wards only in the present Corporate Borough of Ramsgate, within whose jurisdiction the whole of its inhabitants (6,449) are included, with the exception of 390 people in the more distant country part of the Parish. Ramsgate thus contains 23,321 souls (inclusive of 638 sailors) at the last census, is the Metropolis of Thanet, and has a larger population than Margate by nearly 5,000 people.

S. George's Church was consecrated in 1827 and finally cost £32,000. The inhabitants themselves began by subscribing £3,000; and the Church Commissioners presented them with £9,000. The Trinity House gave £1,000 towards the Tower, in consideration of its usefulness as a sea-mark; and £13,000 was lent to the town by the Church Commissioners, and paid off by instalments of £1,000 a year, with interest, for nineteen years. Mr. Hemsley designed the building, but at his death, Mr. Kendall completed it. The contractor was Mr. T. Grundy, to whose memory a window has lately been dedicated.

The exterior of the Church is in the florid Gothic style, and is surmounted with an hexagonal lantern tower, like that of Boston. The Crypt also is worth seeing. The interior was said formerly to contain 2,000 sittings; but the pews, especially in the Galleries, were cramped and narrow; the West Gallery extended over one fourth of the Church; and those on the North and South projected flush with the fronts of the columns. In 1883 a movement for re-arranging the interior was commenced, for cutting back the galleries to more modest dimensions, for widening the seats, bringing down the organ and choir from their West loft to the East end of the floor of the Church, and for building a spacious Vestry. This work was commenced in 1884 under the supervision of Wm. White, Esq., F.S.A., architect, and finished in the Lent of 1885, at a total cost of £2,900; and a peculiar feature in the restoration was that during that time the Church itself was not closed for Divine Service for a single Sunday.

There are nineteen painted windows by some of the most celebrated artists of the time, besides the east window by Willement, which was presented to the first Vicar, the Rev. R. Harvey, in 1851. There is also a beautiful wall painting at the west end, representing the carrying of a saint by angels to Heaven, lately painted by Henry Weigall, Esq., as a gift to the Parish Church. The Pulpit is the gift of C. W. Curtis, Esq., whose grandfather was the first Sir William Curtis, Bart., of Cliff House, Ramsgate, who was one of the original promoters of the building; and the Vestry is in memory of James Webster, Esq., late Deputy of Ramsgate, and also an original Trustee of the Church, who left a legacy of £500 towards the restoration of the Interior. The living is in the gift of the Archbishop of Canterbury, and is totally unendowed. Carrying with it the Chapel of Ease of S. Mary, and the Mission Church of S. Paul, having nearly 12,000 resident population, and requiring four or five Curates at the least, in addition to the Vicar, to discharge its ordinary work, it is not only the largest but one of the poorest benefices in the Diocese. It is the only parish in the town or neighbourhood which has to depend, with its three churches, on pew rents and offertories only, and yet it maintains schools, containing 1700 children, and has a complete system of charity and thrift organisation for the sick and poor, supported entirely by the subscriptions of its worshippers. There are no eleemosynary institutions in Ramsgate, like those of Sandwich and other old towns; and the Vicar has hard work to find the means for keeping all the parish machinery in proper gear.

S. Mary's Church was originally the Chapel of Ease to S. Lawrence; but, by the Act of 1827, it became the Chapel of Ease to Ramsgate, and so it still remains. Being extra-parochial, no subsequent ecclesiastical divisions can affect its position as a Peculiar. It used to have a gallery on all sides, and at the west end a second gallery intended for the powdered footmen of the nobility and gentry who once filled it. It was built by private speculation in 1790 and was in 20 shares, 18 of which were gradually bought up from the original proprietors by the first Vicar of Ramsgate. In 1861 he resigned the chapelry, and nominated to its ministry the Rev. Alfred Whitehead, one of his curates; for, in the eye of the law and by Act of Parliament, it is a distinct benefice. Mr. Whitehead, who is now Rural Dean and Vicar of S. Peter's, purchased the only two shares held by an original proprietor, and subsequently bought eleven shares which were held by the family of the first Vicar, who had deceased. These were put by him in Trust for ever, for the benefit of the Incumbent of S. Mary's for the time being, and the other seven shares were given for the same purpose by the Rev. R. Harvey, the son of the first Vicar.

The Vicar, as patron, has power under the Act to nominate himself as the Minister of the Chapelry; and, with two exceptions for a few years each, he has always been its Incumbent. The pew rents, under present circumstances, scarcely cover the expenses of the worship, which is now musically of a very high and excellent character; and the Clergy therefore have to depend upon the offertory altogether.

St. Paul's Mission Church, King Street, is in the pointed style, and is lined inside with red bricks and Devonshire marble. It consists of chancel, nave, and one aisle. The cost of the Church was nearly £1,400 and was originated and carried out by two Curates of S. George's, the Rev. J. M. Braithwaite, and the Rev. R. Patterson, who are now well known Incumbents in the Diocese, and by the Rev. Canon Elwyn, the late Vicar. The Ecclesiastical Commissioners have now consented to endow S. Paul's with £200 a year, and a Vicarage, *provided* it be enlarged so as to accommodate 600 worshippers. The present Vicar of Ramsgate, the Rev. C. E. S. Woolmer, has bought the adjacent ground for this enlargement, and has promised to present it. The plans have been approved by the Commissioners; Mr. W. G. Osborne is the architect; and £1,100 have been already promised towards the building; but £1,200 more are required before the work can be begun and finished, and a population of 4,000 cut off from the old

Parish. The spacious schools adjoining, used as the Sunday Schools for S. Paul's, and as the Infants' Schools for S. George's and S. Paul's,—were built by Canon Elwyn, aided by his two Curates aforesaid, on a site previously occupied by a well known public house and dancing saloon; and were finally freed from debt in the Incumbency of Rev. C. E. S. Woolner. The Vicar designate of S. Paul's, is Rev. C. E. Eastgate, M.A., of Merton College, Oxford.

CHRIST CHURCH stands in a pleasant situation at the end of the Vale, and was consecrated on August 4th, 1847. It is a handsome structure in the early English style, built of Kentish ragstone, with stone quoins and buttresses, and consists of nave, chancel, side aisles, with porches on the north and south sides, and a tower surmounted by a lofty spire. It was built by subscription, at a cost of about £7,000, and accommodates about 1,000 persons: one-third of the sittings are free. The roof is of oak open work, and the whole has a chaste and handsome appearance. The living is endowed with £1,000 subscription, and is in the gift of trustees.

ST. AUGUSTINE'S CATHOLIC CHURCH is situated on the West Cliff, near the Royal Crescent. This edifice, both in its arrangements and details, presents a strict revival of one of those fine old parochial churches which abound in this country, and consists of a nave, central tower, a deep chancel enclosed with richly carved oak screens, a lady chapel, St. Ethelbert's aisle, and a south transept, part of which is enclosed, and forms the founder's chapel, with an altar dedicated to the honour of St. Lawrence, the patron of the old parish. At the entrance is the Digby Chantry, with its highly finished carved work. All the windows are filled with painted glass. The interior length of the Church is

90 feet, and the extreme width of transept, 60 feet. It was proposed to terminate the tower with a spire about 160 feet in height. The walls are built of flint, and the dressings, as well as the internal walls, of Whitby stone, The style is early decorated. There are cloisters and a school attached to the church. The whole was erected from the designs, and at the cost, of the late talented architect, A. Welby Pugin, the founder of the church.

Adjoining the church is the Grange, a Gothic villa built by the same gentleman. Opposite to this is the MONASTERY, a very exquisite building in the same style, built with flint and enriched with some very fine carving in stone. This building was erected by Mr. W. E. Smith of this town.

TRINITY CHURCH is situated on the Mount Albion Estate, in the parish of St. Lawrence, and was erected by subscription, a piece of ground for the site being given, in addition to a noble donation, by Lady Truro, then Mademoiselle D'Este. The first stone was laid August 1st, 1844, and the church was consecrated for public worship, June 11th, 1845, by the Archbishop of Canterbury. It was built from designs by Messrs. Stephens and Alexander, of Clements Inn, at a cost of £3,000. The external appearance of the building is quiet and unpretending; the interior is ornamented with a handsome stained glass window at the east end, and a smaller one at the end of the aisle. The roof is of oak, varnished. It is fitted with sittings for 770 persons, 300 being free for the use of the poor.

ST. LUKE'S CHURCH, Hollicondane road, is a handsome new building, erected and founded by the Rev. J. B. Whiting and his friends, to meet the spiritual wants of the now fast increasing population of the neighbourhood.

SAILORS' MISSION CHURCH, Military road.

Jewish Synagogue, Hereson.

Chapels.—Wesleyan, Hardres street; Congregational, Meeting street; Baptist, Cavendish street; Mount Zion, Camden road; General Baptist, Farley place; Primitive Methodist, Queen street; Ellington Chapel, Crescent road; and Plymouth Brethren, Guildford Hall.

The Libraries: Mr. Cottew's, High street; Mr. Fuller's, Queen street; and Mr. Wilson's, 36, Harbour street.

Places of Amusement: St. James's Theatre in Broad street is well situated and easy of access, the spirited Lessee sparing no trouble or expense in catering for the visitors to this favourite resort. St. George's Hall (the Head Quarters of 'A' Company, East Kent Rifle Volunteers) is available for Concerts, &c. Sanger's Amphitheatre, High Street. Granville Theatre, at the rear of Granville Hotel. Etablissement, on the Marina, used for Theatricals in the Season.

The Church Literary Institute, is situated in Broad street. It has a good Reading Room, and is well supplied with magazines and books. The subscription is 5s. per annum.

The Albion Club is held at the Old Marine Library in Cliff street, facing the sea. Proposed members are here subject to a ballot. On the table may be found the daily and weekly newspapers, reviews, &c. There is also a capital billiard table in connection with the room, which from its select character is attended by the principal gentry around.

The Ramsgate Tradesmen's Club, held at Goldsmid place, is situated directly over the harbour; it has also a supply of daily and weekly newspapers, together with a billiard room. A very pretty view is obtained from its windows. Members are also here elected by ballot.

East Cliff Lodge, for many years the residence of Sir Moses Montefiore, is situated on the East Cliff, a short distance beyond the Promenade. The caverns at this delightful abode deserve especial notice; they commence in the grounds about thirty feet from the cliff, and extend to the level of the shore, gradually descending, but so arranged as to admit light, and allow of the growth of shrubs, which appear to the visitor extremely beautiful.

Walks. One of the first walks taken by a visitor to Ramsgate will be Pegwell; starting from the end of the West Cliff Promenade, turn to the right, past St. Augustine's Roman Catholic Church, and again to your left, passing the residence of Mrs. Warre, widow of the late M.P. for Ripon, (this, in her early days was the residence of our most gracious Queen) West Cliff terrace, and a pretty little villa, originally called Baron Garrow's Villa, now in the occupation of Sir J. Croker Barrow, you come to Pegwell; here you will find Hotels, with Pleasure Gardens and Marine Walks; in these same Gardens are very snug little summer houses, capable of holding a pair or more, and many there are, who, during the summer time adjourn from the more noisy Ramsgate to indulge in those shrimps for which Pegwell is noted; you may then keep along the lane through Chilton, which brings you into the Canterbury road, and so home through St. Lawrence: or if on arriving at Pegwell you wish to extend your walk, do so by keeping along the cliff side to the Sportsman, which is on the Sandwich road, where every accommodation is provided for the faint and weary traveller. The walk to Pegwell and by Chilton is 3 miles, the one to the Sportsman and back by the carriage road is 4½ miles. Another very nice walk is to St. Lawrence, passing Ellington house. On reaching St. Lawrence Church the visitor must turn to the right past the schools, over the Railway Bridge, and taking the first turn to the right, come out in the Margate road near the Water Works; then returning towards Ramsgate, on the left, in the hollow before ascending to the town is a road which takes you past the South Eastern College, through West Dumpton to Dumpton Grove; the road to Ramsgate is from here straight, but instead of returning through King street, which is not the best entrance to the town, turn across the fields towards East Cliff Lodge, and the town may be entered by Wellington Crescent: the walk is about three miles.

The next walk we would advise, will be along the Margate road, taking the turn to the right at the Waterworks you pass Newlands, then on by Fairfield to St. Peter's, (with its fine old Church), by Upton, Bromstone, and Dumpton, you may return to Ramsgate; this walk, one of the pleasantest in this part of Thanet, is about 4 miles. Then there is the walk by the Sands, at low tide to Dumpton Gap, which is the first opening in the cliff, thence on the top of the cliff to Broadstairs, after seeing which place (so accurately described by Charles Dickens in Bleak House), you can either come home by the high road, in which case you will see nothing of importance that has not been mentioned in previous walks, or returning by the top of the cliff instead of descending to the Sands at Dumpton Gap, passing East Cliff Lodge, and thence past the Granville. A longer walk is to Broadstairs, then by Stone Farm and North Foreland Lodge to the North Foreland,—passing the spot famous for the battle between Duke Wada and Earl Ealchere in 853, and known as Hackemdown Banks,--thence by way of Reading Street to St. Peter's, and by Rumfield Trees into the Margate Road; this walk is about 8 miles. Again to St. Lawrence, turning to the right past the Church, taking the first turning to the left, across the railway, through Manstone to Quex Park and Margate, home by rail: this walk is about 9 miles; or, when at Quex, continue to Birchington and Margate, but this is nearly 12 miles.

RAMSGATE, from its bracing air and situation, as well as from being naturally so well drained, has become a favourite place for the Medical profession to order convalescent patients. A Cottage Home in Belle-vue road, to which those unable to pay the expense of lodgings can be admitted, has been worked for some years by the liberality of the Misses Cotton, and other subscribers whose help they have enlisted. There is also one at Finsbury House. The Sailors' Home Infirmary is in West Cliff road. The Sailors' Mission Room and Church have been erected by Voluntary Contributions, and are situated close to Jacob's Ladder by the side of the west pier, as well as the Smack Boys' Home.

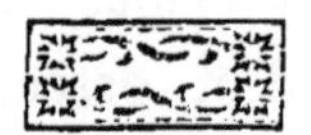

Margate.

MARGATE is par excellence the sea-side resort of City men, and during the height of the season, on the Jetty or Promenades, may be seen almost as much of London as in the most frequented haunts of that crowded mart of the world, many finding it more pleasant, and as economical, during the summer months, to leave their London houses, and so spend Sunday and Monday with their families at the sea-side. It is extremely healthy, as the motto, "Porta maris Portus Salutis," on the Borough Arms signifies. In 1881 it contained 18,236 inhabitants The town is built upon two hills terminating in a valley, up which in days gone by, the sea flowed in the direction of St. Peter's. The Borough of Margate was incorporated by Charter in 1857, and is divided into wards: Marine, Cecil, Pier, and Fort; the Council consists of a Mayor, four Aldermen, and twelve Councillors, each ward being represented by an Alderman and three Councillors. Its antiquity is considerable, having been a member of the Port and Town of Dover from a very remote period; and even in Leland's time there was a pier "for shyppes," but "sore decayed," the time of its erection being unknown. Lewis, in 1723, thus describes Margate: "The ville of Margate is situated partly on the side of a hill, and partly in a little valley, one end of which goes into the sea. It is a small fishing town, irregularly built, and the houses very low, and has formerly been of good repute for the fishing and coasting trade.

The name of the town was probably taken from the water which flowed from the brooks into the sea, down the valley spoken of as originally serving the town; this Hasted appears to have thought was the case on account of the name Meregate, signifying a gate into the sea; many of the older inhabitants can hardly now be brought to call the place after its proper name, but call it Meregate or Mergate. The progress of Margate is, however, evinced from the fact of its having a very beautiful Pier, finished in 1815, at a cost of £100,000; it is 903 feet long, 60 feet wide in the broadest part, and 26 feet high; also the new Jetty, the first pile of which was driven in May, 1853. This elegant and useful erection was constructed by Mr. Samuel Bastow, of Hartlepool, from the designs and under the superintendence of Messrs. Birch, of London, and is a very great benefit as well as ornament to this fashionable sea-port town, being one of the finest marine Promenades in the Kingdom. The head of the Jetty was, during the year 1876, greatly improved, but during the autumn of 1877 a very severe gale occurred, which driving the vessels from the Margate Roads, in consequence of a sudden change of wind, stranded some 10 or 12 large vessels in front of the Marine parade; at the same time a vessel which had been wrecked, and which was on the rocks on the east side of the Jetty, was driven with such force, through the great height of the tide, as to cut the Jetty in two, close to the town, and at the same time to carry away the Refreshment rooms, &c.; many people being on the Jetty at the time were cut off and obliged to remain through this dreadful night, until the the sea receded and daylight broke. The Jetty, during the summer season, is one of the most frequented promenades in Margate. *The Medical Records* says respecting it :—" By its position on a promontory of land, it is surrounded on three sides, viz., the east, north, and south, by the sea. The subsoil is chalk, which, with its geographical situation, renders the town peculiarly airy and bracing, the ground quickly becoming dry after rain. The town is a very open one, and stretches about a mile and a half along the cliff. There are probably few of which the climate is so bracing and invigorating, and a week or two's residence at Margate often changes the jaded and pallid Londoner into a bronzed and robust looking athlete. The bathing is cheap and very safe, but foreshore is flat, and deepens so gradually that the swimmer can get his plunge with more delight and gratification in the deeper water of the south coast. The water supply is good and plentiful ; the water is hard and impregnated with chalk, but it is taken from springs outside the town, and is free from any source of contamination. The lime in the water renders it beneficial in rickets and scrofulous diseases."

THE ROYAL CRESCENT is the principal feature on leaving the Railway Stations. The sea-front commands a magnificent view, comprising, as it does, the Town of Margate, the Harbour, Pier and Jetty, and also a sea view from the Nore to the North Foreland. In front of this Crescent is a Sea Wall with a promenade on the top extending 900 feet. The new Town around Cliftonville, springing up with its terraces of houses, is now fast becoming the most fashionable part.

S. JOHN'S CHURCH, situated at the southern extremity of the town, about half a mile from the pier, is dedicated to St. John the Baptist, and was at first one of the three churches of Minster. It was built 1050, and made parochial in 1210. In 1845 the Church was entirely restored, repaired and decorated, by private subscription. Again in 1876, it was restored, the galleries being taken down, the whole re-seated. It is now a very commodious church. There is an antique font, richly carved, of the time of Henry VII., and in the middle aisle was a tombstone without any inscription, having a cross on it and the Greek X intermixed, which signifies it is for one of the priestly order; perhaps this might be the monument to St. Imarus, who was a monk of Reculver, and is said by Leland to have been buried in this church. Among the ancient memorials and curious brasses is one to Thomas Smyth, dated 1833. On a brass plate is the effigy of a priest, and inscription for Thomas Cardiffe, priest of this church for 55 years, dated 1515. John Daundelyon, Kt., is remembered on a brass plate, dated 1443. There are also memorials for the Norwood, Crisp, Cleve, Cleybrook, and other ancient families of the parish.

TRINITY CHURCH. The first stone of this elegant structure was laid September 28th, 1819, by his Grace the then Archbishop of Canterbury. The church is built with brick cased with bath stone, and is divided into a lofty nave and two side aisles, the whole of which are elaborately groined.

THE CHURCH OF ST. PAUL, Cliftonville, is in the decorated style, built of brick faced with rubble, Kentish rag and Bath stone. It has chancel, nave, and aisles, with tower at the west end, through the lower part of which is the principal entrance. There is accommodation for 800 persons. The total cost, including vicarage, was £9,000. The land was given by T. D. Reeve and W. Andrews. Esqs., of Margate. The architect was R. K. Blessley, Esq., of Eastbourne. The living is a vicarage, annual value

£300, with residence, in the gift of trustees.

THE ROMAN CATHOLIC CHAPEL is in Victoria-road. The WESLEYAN METHODIST CHAPEL is in Hawley-square. ZION CHAPEL, Addington-square, belongs to Lady Huntingdon's connexion. The BAPTIST EBENEZER CHAPEL is near New-cross-street. The INDEPENDENT CHAPEL is in Union-crescent. The CALVINIST CHAPEL is in Love-lane. The PRIMITIVE METHODIST and CHRISTIAN BRETHREN'S CHAPELS are in the Dane.

THE TOWN HALL, built in 1820, forms one side of the Market, with the Police Station beneath. It has upon its walls portraits of F. W. Cobb, Esq., G. Y. Hunter, Esq., and several others, besides a bust of the late gallant Sir Thomas Staines, K.C.B.

THE ROYAL SEA-BATHING INFIRMARY is situated just outside the town on the road to Birchington, and close to the Stations of the South-Eastern, and London, Chatham and Dover Railway Companies. This philanthropic institution was commenced in 1792, when the first stone was laid by Dr. Lettsom, the first projector. It is supported by voluntary contributions, and is under the patronage of her Majesty. The main object of this institution is to bring it within the power of poor and needy scrofulous persons, in all parts of the country, to obtain that relief which Sea-Bathing alone can give, and which the bracing nature of the air here also tends to assist. It has accommodation for more than 200 persons.

THE LITERARY AND SCIENTIFIC INSTITUTION is in Hawley-square, and was established in 1839. A good reading room is always open for the accommodation of the subscribers, as also a library for reference and circulation. The museum of the institution has a very choice collection of British Birds, as well as some of the most curious specimens of Plants indigenous to the Island, and is open for the inspection of visitors.

THE OLD THEATRE, in Hawley-square; built in 1787, has been re-erected, and a very handsome one now fills its place, with accommodation for 2,000 persons, under the managment of Miss Sarah Thorne.

THE GROTTO is situated in the Dane. It is very curiously cut out in the chalk, and prettily and tastefully covered with shells. This beautiful specimen comprises upwards of 1,850 feet, executed in the most regular and elaborate designs. It is exhibited to the public at a small charge, and is well worthy the attention of the connoisseur.

THE POST OFFICE is in Cecil-square, opposite the Assembly Rooms.

There are several good Libraries situated in the High-street.

Westgate-on-Sea.

VISITORS to Margate for some years will remember the time when the only place termed Westgate was a Coast-guard station with a few small houses, and not far off the celebrated little house of call, the *Hussar*, at Garlinge; now there has sprung up one of the most charming resorts for health seekers. The land having been bought by a Company, and well advertised as the Westgate-on-Sea Estate, and being highly recommended by several eminent Physicians, particularly as a resort for all persons suffering from cutaneous diseases, its popularity has spread so rapidly that it bids fair soon to rival its larger neighbour of Margate. The Estate is perfectly laid out, and the stipulations are such as to ensure a good class of house; and, if we may judge from the way in which it has lately improved, will soon become a most fashionable watering place. A very handsome new Church has been erected in the hamlet of Garlinge. The Granville express Trains, to and from London, call at the Station of the London, Chatham and Dover line which runs close to the Estate.

Broadstairs.

BROADSTAIRS or BRADSTOW, (the Saxon for Broad Gate) so called originally from the broadness of the gate or way into the sea, has had its praises sung in inimitable fashion by Charles Dickens. It is curious that, situated between two such bustling and crowded places as Margate and Ramsgate, Broadstairs should have retained so long, its old-world appearance and old-fashioned ways; but that it has done so must be a source of unbounded satisfaction to those for whom a fashionable watering-place is an infliction, and roar or bustle, an agony. For people who want absolute quiet, combined with bracing and healthful air, Broadstairs is the watering-place most within reach. The climatic characteristics of the place are the same as at the other resorts in the Isle of Thanet, and need not therefore be dilated upon. The subsoil is of course chalk. It is a favourite resort for gentry, and those who require quietness will certainly find the united advantages of tranquility and seclusion here; and as the bathing is good, it is hardly to be wondered at, that from a small hamlet in the parish of St. Peter's, it has lately risen to be a favourite watering-place, and is now dignified with a Local Board. The Church is dedicated to the Holy Trinity, and like so many other buildings in this neighbourhood, is faced with flint; it was erected in 1822. Its Curacy, valued at £180, is in the gift of the Vicar of St. Peter's. About the time of Henry VIII. a small wooden pier was erected here, to protect the fishing craft, near to which was the Chapel, dedicated to the Virgin Mary, wherein was her image, called our "Lady of Bradstow;" the sailors in passing this place used to lower their topsails in order to salute it, so great was their veneration for their Lady saint. Whales have been stranded in this neighbourhood several times within the memory of many persons living.

St. Peter's.

THIS delightful little village will well repay a visit, whether made by the Archæologist or the mere casual observer. Here although so near to the bracing sea air, you find yourself in a quiet country village, with its neatly trimmed lawns variegated with flowers; the luxuriant crops shewing that the farmers here, as well as having good land, know how to till it. Near to the village are many gentlemen's houses, in fact the air is that of a thriving well-to-do population. The village is nicely surrounded by trees, which are rather rare in the Isle of Thanet. It is evident St. Peter's must have been a place of some note in days gone by, for Hasted says, "It is an antient member of the Town and Port of Dovor." In 1563 there were within the parish one hundred and eighty-six households; there are also within the parish several hamlets: Upton, Bromstone, Dumpton, Stone, and other outlying places, such as Sackett's Hill, Reading Street, Baird's Hill, Callis Court and Dane Court; this latter was a gentleman's seat in former times, and gave to its holder the family name of Dane; but we find from the history of it, that the law of gavel kind, which rules in this part of Kent, divided the estate, and so it passed into many different hands, the name being lost. The Church, as the name of the Parish implies, is dedicated to St. Peter; the living is in the gift of the Archbishop of Canterbury. The Church is built, as are nearly all in this neighbourhood, of flint, and the windows and doors cased with stone; the tower acting as a landmark for seamen running past the Goodwin Sands; it was built in the year 1184, and restored in 1853. It is now perhaps, one of the most beautiful churches in this part of Kent. There are many brasses and tablets for the curious to decipher. In 1730 a Mrs. Elizabeth Lovejoy left a sum of money for the repairing of the Chancel, and keeping in order her own and her husband's monument which was therein; as she died of apoplexy, before executing the will, it caused many lawsuits, but was eventually proved to be binding. In this church are buried the remains of Thomas Sheridan, father of the famous Richard Brinsley Sheridan, of whom it may be said, he combined the remarkable power of the orator with the keenness and perception of the wit; and although Sir W. Temple says "none was ever a great poet that applied himself much to anything else" we certainly find the good, beautiful, and highly imaginative poet, as well as the dramatist, in the son of him who here reposes. A noble tablet was here erected in memory of Mr. Sheridan, by an old friend of his, a stranger and visitor to this neighbourhood, who, finding that his remains rested here without such tablet or memorial, ordered one and had it fixed at his own expense. This friend is himself buried in the north aisle of the church.

The Kentish Sampson, Richard Joy, was born and buried here: the extraordinary records of his strength—breaking a rope that would lift 35 cwt., and lifting from the ground 2,240 lbs., would certainly lead us to the supposition that we are sadly degenerate, and that our forefathers were men of altogether a different calibre.

The village of St. Peter's, although still so popular amongst visitors, has not the inducement which it held out some years since. The Gardens, laid out with great taste in 1818, and which for so many years were well attended—even as many as 800 sitting down to breakfast here—have now disappeared, and on the ground has been erected the National School, for which purpose it was purchased by the Rev. I. Hodgson.

St. Peter's Orphanage, founded by Mrs. Tait, wife of the late Archbishop of Canterbury, is a convalescent home for destitute orphans and convalescents requiring sea air, and is capable of accommodating 60 children, and 20 convalescents, who pay 7/- per week during residence. The orphan children are nominated by ladies, who pay £12 a year for each, and undertake to watch over and befriend the orphan in after life; the girls receive education and industrial training. It is built on ground given by the late Archbishop, and is a stately and substantial structure, of early Decorated Gothic architecture, from the designs of Mr. John P. Seddon; it is faced with flintwork, with freestone dressings, and relieved with red brick diapering: a chapel occupies the space of two upper storeys in one wing, and the polygonal apse of this, forms the principal feature externally; schoolrooms, a refectory, and sisters' common room, occupy the ground floor, and dormitories the remainder of the two storeys over; and in a basement under the whole are laundries, kitchen, play-room, &c.

The North Foreland.

A MILE and a half north-eastward from St. Peter's Church, and near to the edge of the cliff, is situate the NORTH FORELAND LIGHTHOUSE, so called in contradistinction to the South Foreland, midmay between Deal and Dover: there is no doubt that this was the Cantium of Ptolemy, and the Cantium Promontorium so well known to the Roman seamen. It would be needless to endeavour to show cause for this beacon, placed in such a situation, with a dangerous point of land jutting out into the sea on the one hand, and on the other the far-famed Goodwin Sands, the north light of which appears opposite this; suffice it to say that, with even these two lights to guide him, the seaman is often driven ashore at this point, not being able to get into the Margate Roads on the approach of dirty weather. About the end of the last century it was rebuilt and thoroughly remodelled, and, from being hired, the same as a toll-gate, it was bequeathed to Greenwich hospital, and afterwards tranferred by Government to the Elder Brethren of the Trinity House. This lighthouse is now constantly visited in summer time by those who wish to get a commanding view of the surrounding scenery, and the obliging person in charge willingly gives any information in his power respecting the lights, which are very powerful, and may be seen in clear weather at the Nore 30 miles distant.

Near to the lighthouse, and between it and Kingsgate, are Hackemdown Banks, where formerly were tumuli, which, tradition said, contained the bones of the English and Danes who were killed here in the battle between Earl Ealchere and Duke Wada, in 853. It seems strange that a battle should be fought so close to the edge of the cliff, but as we know, formerly the land stretched here out further into the sea, at that distant time it might not appear so bad a position; but Asserius Menevensis tells us that in consequence of its being fought so close to the sea, a great many were pushed into it and drowned; and, until very lately, a gate into the sea near this place was called Battle'em Gate by the inhabitants of this part of the Island. One of the tumuli was opened May 23rd, 1745, by Mr. T. Read, in presence of several hundred spectators, when, a little below the surface of the ground, numerous graves were found dug out of the chalk, and covered with flat stones.

Kingsgate.

AT KINGSGATE as it is *now* called, Bartholomew's Gate as it *was*, we find the annexed Latin distich, affixed to the Gate, by a Mr. Toddy, of Joss, after King Charles II. and his brother, the Duke of York, landed here, on the 30th June, 1683, on his way from London to Dover:

"Olim Porta fui Patroni Bartholomœi
Nunc, Regis jussu, Regia porta vocor.
Hic exscenderunt. Car: II. R.
Et Ia. dux Ebor, 30 Iunii, 1683."

In former times there was nothing but fishermen's huts at this place: but Henry Lord Holland coming to reside here, bought ground of Robert Whitfield, Esq., for a mansion, which was built on the model of Tully's Formian villa on the Coast of Baiœ, but as it has gone to decay altogether, it is useless giving a detailed account of its proportions and architectural beauties. Kingsgate is at the present time a very popular drive with both Ramsgate and Margate visitors, the distance being short and the drive nicely varied, going by the cliff and returning further inland. KINGSGATE CASTLE is a very picturesque old building covered with ivy, standing upon the cliff, which seems as though it must soon be undermined from the constant wearing away of the chalk by the action of the sea; a commanding situation and fine sea view makes it a delightful summer residence. Artists will also find this a charming place for making sketches. The *Captain Digby*, a small hostelry facing the sea, on one flank, and Kingsgate Castle on the other, gives every inducement to the weary traveller to stay and enjoy the bracing winds from the North Sea.

Dandelyon.

DANDELYON or DENT DE LYON, so called from a family of the latter name living here so far back as Edward I., is a pretty rural spot, but is only worthy of remark as having an old gate in good state of preservation, which appears to be part of a system of defence to this seat; judging from the gate-house having loopholes and battlements, and the strength with which it is built, it would stand a long siege. Over the gate appears the armorial bearings of the Daundelyons. Under the right of the gatehouse, in 1703, was found a large room, containing many lachrymatory urns. Some years since, when travelling was not so easy, and races were less common, a meeting was held here in September, at which we have very good authority for stating, all the gentry of this part of Kent used to attend. Dandelyon is about one mile west of Margate.

Birchington.

BIRCHINGTON, which is on the direct road from Margate to Canterbury, is about three miles north of Minster, and the same distance west of Margate. It is a member of the Town and Port of Dover, and in 1711 it was enacted that Margate, St. Peter's, and Birchington should be deemed a distinct division of that Liberty. In the Church are many monuments, brasses, &c., to the families of the Crispes of Quex, and others. The spire of the Church serves as a landmark for vessels steering from the Thames round the North Foreland.

GORE END, in this parish, was formerly a place of great note, in fact, it is supposed that the Church originally stood here, but being lost by the falling away of the cliff, was erected at Birchington.

At the north-east of this parish is WESTGATE, from which place it is said Dompneva's deer commenced its course.—*See Legend to that effect under the description of Minster.*

QUEKES or QUEX is in the south-eastern part of this parish, in a line with Manstone. The seat originally belonged to a family of the same name, and afterwards to that of the Crispes. An amusing anecdote is told of one of that name who was appointed sheriff during the commonwealth, and whom some staunch Royalist surprised in his bed, and forcibly carried to Ostend and then to Bruges, as prisoner. Cromwell being applied to, suspected it to be a collusion to obtain £3,000 for the use of Charles II., and so issued an order in Council that he should not be ransomed. It seems strange that he was so easily conveyed away, especially as he was aware of a determination to carry him off, and had secured his mansion by hospitality, receiving his neighbours into his house as a guard. It appears the nephew, Thomas, was forced to sell some of the land to obtain money for his release, which he effected about eight months after his being taken away. It is said of him that, although on the continent so long, he could not conquer more of the French language than *bonjour*, hence his appellation, Bonjour Crispe.

Reculvers.

RECULVERS or REGULBIUM is another spot of great historic interest, although from the inroads of the sea, there is great fear that it will soon be entirely carried away, as a large portion of it has already been. The waters here occasionally open to view some grave, and the visitor may often see the bones of some of our Saxon ancestors unearthed by the waves. It was here the Wantsume flowed into the sea. The south and east walls of this old ruin are still standing, they appear to be about 12 feet thick, and are built of flint and pebbles ; in Leland's time it stood about half a mile from the sea ; the Saxons called this, Raculf Ceastre. It was here Ethelbert retired after his baptism and conversion by St Augustine, and built himself a palace, so that the ruins we now see here, although ancient are not nearly so old as Rutupinum ; there we have the remains of the old Roman fortress, here only one built on the site of the Roman one. It seems that the towers must have been washed away ere this, but for the Trinity House stepping in and putting down groins in front to protect them from the sea, as they were of great use to mariners ; but soon after the spires were blown down, and in place of stone the present (apparently of dark or stained wood) were erected. The sea wall extends from the Coast Guard Station here nearly to Birchington, which prevents the sea encroaching as it previously did. You may see what remains of the River Wantsume, as here it discharged at its north mouth and drained the flat lands adjacent. Visitors to Ramsgate or Margate will find a pleasant day's recreation by taking train to Grove Ferry and footing it across the meadows to this interesting relic of by-gone days.

Of late years Reculver Castle has been explored by Mr. Roach Smith, whose investigations have thrown a new light upon the inquiry. The work is manifestly Roman: the chancel arch was triple, resting upon two columns, and they were of Roman brick. It has been asked by an able critical writer, "Is it a church built out of some Roman building, which, even in its ruined state, was capable of being adapted to such a purpose; or is it simply a church built, after the conversion of the Kentishmen, by the Roman missionaries in the Roman manner? The work, though Roman, cannot be called classic. It may be work of the very latest Roman days, or even of Welshmen left to their own skill after Honorius had withdrawn his legions. Or it may be the work of the earliest Christian Englishmen and their instructors. In either case, it bears witness to no continuous Roman traditions, such as meet the inquirer at every step of a journey through a Romance-speaking land."

St. Nicholas at Wade.

ST. NICHOLAS AT WADE is about two miles from Monkton; its name is from the saint to whom its church is dedicated, and its near proximity to the Wantsume, which about the time of the building of this church became so shallow as to be able to be waded through; this was about the year 1200, at which time it was only a chapel to Reculver, but in the year 1300 a vicar was appointed to it; the patron is the Archbishop of Canterbury.

Sarre.

SARRE is about a mile from St. Nicholas. It is now only a small village, but formerly, from being the principal ferry to this island, and a place where most of the shipping anchored on their way to the North Mouth, was then much more populous. It is by some writers supposed that the original name of the water which ran by here from the North Mouth was called Serre, hence the name of this village; but there is no doubt that in Bede's time it was called Vantsumus or Vantsume. The bridge over a small stream just outside Sarre is supposed to be the position where at that time two ferry boats were kept constantly in use to convey men, &c., across: the toll was granted by Egbert to the Abbey of Minster. At that time there was a parish church to this place, but in 1540 we find that Divine service was discontinued here.

Monkton.

MONKTON is about a mile from Minster, and two from Sarre; its name seems to have been given to it from its belonging to the monks of Christ Church, Canterbury, to whom it was given by Queen Ediva. Archbishop Stratford, in the reign of Edward III., obtained the grant of a market to be held here on Monday, and a fair on the Nativity of the Blessed Virgin. The Church is dedicated to St. Mary Magdalen; the Archbishop of Canterbury is patron. According to Lewis, the following couplet was to be read at the west end of this church:—

> "Insula rotunda Tanatos quam circuit unda,
> Fertilis et munda, nulli est in orbe secunda."

This will shew the estimation in which this Island was held

for fertility. MONKTON COURT and CLEVE COURT are both in this parish.

Minster.

MINSTER is on the South-Eastern Railway, about five miles from Ramsgate, and was originally so called on account of the Monastery which was here. The Church, dedicated to St. Mary, is very handsome, consisting of a nave and two side aisles, a cross sept, and east chancel; the architecture of the nave is Saxon, but the transept as well as chancel are Gothic.

St. Mildred's Abbey which, according to the best information we can gather, was within a short distance of the Church, must have been in those days connected, as a wall three feet thick, near to the present Vicarage house, has been traced down in the direction of the Church, and exactly agrees with one inside the same, this view was taken by Thomas of Elmham. In the chancel are some curious carvings, and upon one of the Bells are inscriptions, an effigy, and also a monogram. Sir Stephen Glynne says: "one of the very finest Churches in the country. The tower and belfry loft are worth devoting a short time to visit, as it appears probable this was used in bygone days as a watch tower, as from here could be well distinguished the warning lights of Richborough, North Foreland, the Harbour which then lay where Ebbs Fleet is now, and the ferry crossing to the main land at Sarre and doubtless signals passed between this tower and the one at Monkton from which Reculvers light could be seen.

The legend connected with the Monastery here in 720 is very curious. Ermeured, King of Kent, dying, left two sons and four daughters to the care of his brother, who in like manner dying left his son Egbert to fulfil his brother's will; and Thorne says: "he, to secure the kingdom to himself, had the two princes put to death, and buried, to prevent any discovery, under the Royal Throne, but which was revealed by a light from heaven pointing out the place. At this Egbert was very frightened, and St. Theodore, Archbishop of Canterbury, prevailed on him to send to Dompneva, one of the sisters of the murdered Princes, to ask her pardon, and to make restitution. But Dompneva would none of his rich presents, but forgave him, and requested a piece of land where she might build a Monastery, and there, with the Virgins, her daughters, who with her were devoted to God, might continue to pray to Him to pardon and forgive the king for the murder of her two brothers. Egbert granting this petition, asked her how much land she would have to endow it, upon which she said, 'as much as my deer can run over at one course.' This being granted, the deer was let loose at Westgate, in the parish of Birchington. While they were following her, Tunor, the king's lieutenant, who had murdered the Princes, cried out that Dompneva was a witch, and the king a fool, he therefore tried to stop the deer by crossing her, but whilst he was so doing the earth opened and swallowed him up, on which the king was very much afraid, and delivered up to Dompneva the whole tract of land that the deer had run over. Mildred succeeded her Mother as Abbess, and to her Edburga, who removed the body of St. Mildred from St. Mary's into a temple which she built for the accommodation of more virgins who had joined the order. This St. Edburga died 751, and very shortly after the Danes commenced their visits to the island, plundering and laying waste, and of course the Monastery of St. Mildred was not excepted. Being constantly subject to these depredations, the sisters fell away, and somewhere about the year 800 the Danish army entirely destroyed the Monastery of St. Mildred." There are many other legends concerning St. Mildred to be found narrated in the old Historians, Lewis, Hasted, &c.

Adjoining the churchyard is MINSTER COURT and ABBEY, which was fitted up by the monks of St. Augustine as a residence for those to whom the care of the manor was entrusted. An attempt was made to destroy it, in 1318, by their tenants, who, enraged at the levies and distresses issued by the abbot, gathered together to the number of 600, armed with bows and arrows, swords and staves, set fire to the gates, and kept the monks imprisoned for the space of fifteen days.

On the high ground in this parish, close to the Canterbury road, is one of the finest views in this part of Kent; on the one hand the spires of Reculver, Quex, and Birchington, Island of Sheppy, mouth of the Thames, and Coast of Essex; and on the other the Downs, the towns of Deal, Sandwich, and Ash, the spires of several Churches in East Kent, Ruins of Richborough Castle, Canterbury Cathedral, and in very fine weather the cliffs of Calais.

Ebbs Fleet.

EBBS FLEET is a little to the right of the Sandwich road; there, as before mentioned, it is supposed Hengist and Horsa landed, and also St. Augustine. It appears this was the common landing place on this side the country, and seeing that St. Augustine and his Monks came overland through Gaul, nothing was more natural than that they should make for this port; and as we believe it is hardly doubted that it was the king of Kent, who met them on their arrival, none other so favourite a spot in Kent could be found. One of our poets speaks of the landing as certainly taking place in Tenet, and Stanley, in his Historical Memorials of Canterbury, says, "Ebbe's Fleet is still the name of a farm-house on a strip of high ground rising out of Minster marsh, which can be distinguished from a distance by its line of trees, and on a near approach you see at a glance it must once have been a headland or promontory running out into the sea between the two inlets of the estuary of the Stour on one side, and Pegwell Bay on the other. What are now the broad green fields were then the waters of the sea. The tradition, that "some landing" took place there, is still preserved at the farm, and the field of clover which rises immediately on its north side is shown as the spot.

Stonar.

STONAR can hardly be spoken of as one of the places of note in this island, as there are now no remains of it; but in consequence of the great interest attached to it as the Lapis Tutuli of the Romans, as well as from its position as the key to Richborough, we will point out its original site, about midway between Ebbs Fleet and Sandwich: it was at one time separated from the main land, and completely insulated, but the Wantsume getting choked up, at this its mouth, it was thought possible to build walls, and so prevent the sea at spring tides overflowing the land, which being done, the wide bed of the river gradually dried up, and in process of time became arable land.

Here, Turkill the Dane fought with the English, on his coming over with his fleet; Vortimer's battle with the Saxons was also close to the spot, after which he drove them into their highland fastnesses. It is supposed Austin, the Monk landed here on his passage from France. In 1015, King Canute, having murdered his brother, and defeated Edmund Ironsides, to expiate his crime and ingrate himself with the Clergy, granted the manor of Stonar to the Abbey of St. Austin, and to the Priory of Christ Church.

Thorne, in describing the ruins of this once noted town, tells us that, in the year 1385, Simon Burleigh, Constable of Dover Castle having invited the French to invade this country, made overtures to the Prior and Chapter of Trinity Church, Canterbury, to remove Thomas á Becket's Shrine and their treasure to Dover, which was only a ruse to get possession of their gold. On his being disappointed, he at once got eighteen galleys (part of the French fleet which was to invade the country) and plundered and burnt down the town of Stonar; but the Abbot of St. Austin, marching into the Island by Sarre, came up against them and made them retire to their ships, but not until they had completely laid waste and pillaged the town.

Richborough Castle.

STANDING within the ruins of this remnant of the past, now so lowly, once so great, we cannot but be reminded of the changeableness of all earthly things, and how the high looks of the proud may be brought low. Surely it cannot be strange that there should be much dispute as to the construction of this immense fortress, for on the one side we have a long wall without a loophole of any sort, and this on a side that would be expected to be fortified. What was the shape of the building? How had these ruins come here apparently below the foundation of the other part of the Castle? Was there a wall facing the estuary? Or, in what way were inmates secured from attacks by sea? these are all questions one asks oneself on viewing these ivy-clad ruins. The Romans principally applied the name Rutupiæ to this fort, but it was evident that was meant for Reculver and this in conjunction, which are always termed the Rutupiæ. This is by most writers termed Portus Rutupinus, from its situation as the port at the entrance to the Estuary, where we find was a bay nearly five miles broad. It has been contended that Julius Cæsar landed here; but Dr. Halley, in his Philosophical Transactions, No. 193, contradicts and refutes that supposition, aud plainly proves that it was at Deal. The best description of the Town and Castle is given by Leland, in his Itinerary, written during the reign of Henry VIII., and these are his words: "Ratesburg otherwyse Richeboro was, or ever the ryver of Sture dyd turn his botem or old canale, withyn the Isle of Thanet, and by lykelyhood the mayn se came to the very foote of the castel. The mayn se ys now of a myle by reason of wose, that has there swollen up. The scite of the town, or castel, ys wonderful fair upon a hille. The walles the whych remayn ther yet be in compase almost as much as the Tower of London. They have been very hye, thykke, stronge, and wel embateled. The mater of them is flynt, marvellus and long brykes, both white and redde, after the Briton's fascion. The sement was made of se sand and smaul pible. There is a great lykelyhod that the godly hil abowte the castel, and especially to Sandwich ward, hath bene wel inhabited. Corn groweth on the hille yn marvellus plenty; and in going to plowgh ther hath owt of mynd bene fownd, and now is, mo antiquities of Romayne money than yn any place els of England; surely reason speketh that this should be Rutupinum.

The walls as they now stand are, the northern 440 feet long, and the southern about 260 feet, being from 10 to 30 feet high, and 11 to 12 feet thick; it is built in the usual style of Roman masonry, alternate courses of square stones and red and yellow tiles. There are, also, the remains of a square tower near to the centre, and other towers in the walls. Near the middle of the walls may still be seen, when the corn is not growing, the remains of a cross, or base of a building in that form; its shaft is 87 feet, and arms 46 feet. Beneath this, in 1822, was discovered a subterranean building; the traditional name of this is St. Augustine's Cross, as, after St. Augustine's meeting with Ethelbert, he sojourned for some time here, and it is surmised that it was built at that time, probably in commemoration of his planting the cross in these dominions. The Latin poets occasionally refer to it, but more especially Juvenal, who seems to think the flavour of its oysters was such as none other could compare with; it is narrated that, so fine were those taken here, the Romans could tell whether they were from Richborough or not by their taste. Lord Granville has erected a Cross in commemoration of the landing of St. Augustine, on land near Ebbs Fleet, which will well repay a visit.

During the Summer Season Trip Trains at Cheap Fares run between Ramsgate and Margate, Canterbury, Deal, Sandwich for Richborough Castle, Dover, and Hastings, &c., a record of which is kept in Wilson's Visitors' List and Directory, price one penny, published every Saturday.

ENVIRONS OF THANET.

Sandwich.

SANDWICH, originally, was one of, if not the, chief Cinque Port, having as its members, Fordwich, Reculver, Sarre, Stonar, and Deal. It is evident that this port came into more notice and greater importance in consequence of the decay of Richborough; but that the present harbour was the original one is, by the old writers, completely set at rest, as they distinctly tell us where now the town is was originally covered with water, so we may conclude the site of the port itself was, at the subsiding of the waters in these parts, reclaimed from the sea. It was called Lundennac, from its guarding the entrance to the port of London, which is very good evidence that the road to London by the North Foreland was at that time, if a sea way at all, at any rate not frequently used. This name it retained during the reign of the Saxons, but when the Danes supplanted them, they called it, from its sandy situation, Sandwic. In the year 665 or 666 we first read of this place as a port. Stephanus, in his Life of St. Wilfrid, Archbishop of York, mentions that he with his company happily and pleasantly arrived in this harbour. During the Danish incursions this town, from its situation, became the scene of so many engagements, etc., that it was called the most famous of all the British Ports.

In 979 Etheldred gave this town to Christ Church, Canterbury; and Canute, after finishing the building of it in 1203, gave it also for the support of the above Church and the Monks therein; the former having giving it as the land of his inheritance, the latter as the land of his conquest.

This town appears to have been nearly burnt down in 1217 by the French, but was immediately re-built, by royal order, in consideration of the services continually rendered to the Country at large by the shipping of the Port. So great estimation was it held in that, from the Conquest to the time of Richard II., it was frequently visited by royalty, who embarked here for France. The royal Fleets were also constantly at this place, and, in time of need, the 1,500 seamen residing here, manned fifteen ships of war, to the charge of the town. This so enraged the French that, in the time of Henry VI., they were continually committing depredations upon the people of Sandwich, which were brought to a climax by Charles VIII. of France despatching 4,000 men, who landed at night, and, after a very hard fought and sanguinary battle, succeeded in gaining possession of the town, which they fired and the inhabitants they principally butchered; and again not long after, it was ransacked by the Earl of Warwick. Edward IV., to guard against future attacks of this kind, had the place fortified, and the moat dug around it; this, and its position as a harbour for vessels in danger of the Goodwin Sands, caused it again to be prosperous; but after this time, in the reign of Henry VII., when the Wantsume began to decay, and the waters all around to fall off, this state of prosperity was in a measure stopped. Leland speaks of the foundering of a large vessel in the haven in Pope Paul IV.'s time, which prevented the free current of the waters; at any rate several endeavours were made (by Royal Commissions to enquire into the state of the harbour) to save the Port from destruction, and restore it to its original position, but whether for want of capital, or because the Commissioners could see no chance of restoring it, it fell away, and in Queen Elizabeth's time was useless except for vessels of small burthen; but, although as a Port it lost its importance, in the same Queen's reign it became a most thriving place by the establishment of many manufactures by foreigners driven here by the religious persecutions of Brabant and Flanders, which continued until the reign of James I., in which monarch's time it again fell off.

Many interesting accounts are given of the visits of Royalty to Sandwich, etc., but more especially of the one by Queen Elizabeth. This town is noted for its many eleemosynary institutions. In 1722 Henry Corfield founded here a Carmelite Priory called the Whitefriars, which, however, was much increased by Lord Clinton during Edward I.'s reign, and became a most flourishing institution. St. Thomas was founded in honour of Thomas á Becket, by Thomas Ellis, a draper, of this town.

Grove Ferry.

GROVE FERRY, situated about ten miles from Ramsgate, on the South-Eastern line of rail; as its name denotes, is a ferry across the river Stour. It is remakable for nothing but its pretty little Inn, situated by the side of the river, and which, in the season, is visited by numerous pleasure parties, the bowling green attracting some, but more being pleased with the seclusion of the spot and the pretty walks in the neighbourhood. The worthy host deserves every commendation for his endeavours to satisfy the wants of his many patrons, and those who once pay a visit here will certainly feel inclined to renew it, especially if they come during the strawberry season, when they may have some of the finest ever tasted, with cream equally good. Pleasure boats are kept here for fishing excursions. Intending visitors would do well to write beforehand when they intend to spend the day at this pleasant spot, as it is situated so far from any town where fish, etc., can be obtained.

Canterbury.

IT is useless to attempt to give a guide to this city in the space we have allotted to it, as good ones may be obtained by visitors who require them; nevertheless, as some may not care to do so, we give the principal features of interest, with some little recognition of those occurrences which have made them famous. The CATHEDRAL is of course the most interesting object, the Gothic pile of which will well repay more than a passing glance. The great central or Bell Harry Tower is a most beautiful specimen of the pointed style of architecture. The west Towers, which have been recently restored, are exceedingly grand, the only drawback to their view being the houses which surround them, and which prevent a west view of the Cathedral. Part of this stupendous mass of buildings was erected by St. Augustine, but being several times injured by fire, it was found difficult to provide means for rebuilding and restoring it until about four years after the death of Thomas à Becket, when the numerous pilgrims, who came to visit the scene of his martyrdom, brought such gifts as to enable the Monks to do more than they had originally designed, and they consequently began a chapel in honour of St. Thomas the Martyr. Louis VII., of France, came here as a pilgrim about the year 1179, with a train of nobility, etc., and their oblations of gold and silver were very great, the king giving a cup of gold and a royal precious stone, as well as a grant of 100 muids of wine for ever for the convent. It appears that oblations were continued, and consequently building and beautifying, until the time of Henry VIII., who put a stop to them and seized upon the treasures. In after years some of Cromwell's troops turned the Cathedral into a stable for their horses. If the exterior of the Cathedral is grand, the interior is equally so: the Choir, the largest in England, is very remarkable for its beauty; the pillars, alternately circular and octagonal, give a decided effect; the stonework screen at the entrance from the Nave is a very delicate specimen of florid Gothic architecture; the Archbishop's Throne, in the Choir, is also a beautiful piece of carving in stone. On entering the Choir, looking east, on the left are seen the monuments of Archbishops Chichele and Howley, also Cardinal Bourchier; beyond is the high altar of Trinity Chapel and Becket's crown, on the right is the Archbishop's throne, and beyond that monuments to Archbishops Kempe, Stratford, and Sudbury, the latter murdered by Cade's mob on Tower hill. The Nave is a beautiful specimen of pointed architecture, the groining of the roof being well worthy of notice; the west window is very ancient, but well preserved. In the north aisle of the Nave is a monument to Dean Lyall, also one to the 50th Foot, and nearer to the Choir is one erected by the gallant 16th Lancers to their fellow officers and comrades who fell at Aliwal and Sobraon. In the south aisle of the Nave is one to Bishop Coleridge, also one to the 6th Foot, of Jellalabad renown, over the tattered colours of the regiment. The north wing of the Nave is called the Martyrdom; at the west side is the entrance to the Cloisters, which was the scene of Becket's murder. The south wing of the Nave contains the Warriors' Chapel, erected by Edward the Black Prince, in memory of his followers, who, with him, fell into an ambuscade, and were cut to pieces, he himself narrowly escaping. Trinity Chapel is, of course, one of the principal sights in the Cathedral, containing as it does the shrine to Thomas á Becket, monuments to Edward the Black Prince, Henry IV. and his Queen, Joan of Navarre, a brick sarcophagus to Odo Coligny, the Huguenot Bishop elect of Beauvais, and Cardinal Chastillon, who was poisoned by one of his servants on account of his conversion to Protestantism.

KING'S SCHOOL is situated in a quadrangle to the north of the Cathedral; it is worthy of note as containing a staircase which is one of the finest specimens of Norman architecture in the country.

ST. AUGUSTINE'S MONASTERY, supposed to have been built by King Ethelbert as a Cemetery, at the instigation of St. Augustine, stands on the road from Burgate to Richborough, it being the custom anciently to bury the dead by the roadway; in after days it became a flourishing institution, in fact so much was it thought of, that Henry VIII. appointed it a Royal Palace. The Gateway, called St. Augustine's, and the Monastery, are very beautiful, and well worth a visit, although they are somewhat out of the way for visitors to the Cathedral; this also applies to ST. MARTIN'S CHURCH, the first Christian Church in this Country, being the one used by St. Augustine and his Missionaries on their arrival in these parts, and where Ethelbert was baptised; Stanley says, "in St. Martin's they worshipped, and no doubt the mere splendour and strangeness of the Roman ritual produced an instant effect on the rude barbarian mind. And now came the turning point of their whole mission, the baptism of Ethelbert. It was, unless we accept the conversion of Clovis, the most important baptism that the world had seen since that of Constantine. We know the day—it was the Feast of Whit-Sunday—on the 2nd of June, in the year of our Lord 597. Unfortunately we do not with certainty know the place. The only authorities of that early age tell us merely that he was baptised, without specifying any particular spot. Still, as St. Martin's Church is described as the scene of Augustine's ministrations, and among other points, of his administrations of baptism, it is in the highest degree probable that the local tradition is correct;" it is supposed to have been built somewhere about the year, 187. Here, it is imagined, the good Queen Bertha was buried, to whom in a measure we must give the credit of inducing King Ethelbert to embrace the Christian Religion. A sarcophagus in the hall is shewn as the tomb, and the font is also supposed to be the one used for the baptism of King Ethelbert. Some few years since the interior of this interesting old Church was restored by the Hon. Daniel Finch, who added an oaken gate at the entrance to the churchyard.

Among the ruins of ancient buildings at Canterbury, on the

south-west side of the City, near the entrance from Ashford, are the walls of a Castle, supposed to have been built by William the Conqueror; larger than that of Rochester, being 88 feet by 80 in dimension. These remains appear to have been the keep, or donjon, of a fortress, within which it stood, and of which the bounds may still be traced, like those of the Castles at Dover, Rochester, and the White Tower of London, the building being much in the same style with those just mentioned. The original portal was on the north side, and the state chambers on the third story, where alone are found large windows. The principal room in the centre of the edifice was 60 feet by 30; two others on the southern side were each 28 feet by 15; and one on the northern side was 20 feet by 15. In the latter end of the reign of James I. the Castle was granted away from the Crown, and became private property.

Appended is a Table of a few of the most remarkable events connected with Canterbury, arranged in Chronological order, extracted from Murray's Handbook and Dean Stanley's Canterbury.

570 Ethelbert, King of Kent, married Bertha, a Christian Princess, and daughter of Charibert, King of France, who came to England attended by her Christian Chaplain, Bishop Luidard.

596 St. Augustine, sent by Pope Gregory to attempt the Conversion of the English, landed at Ebbe's-fleet, Thanet.

597 Ethelbert embraced the Christian faith. Augustine became first Archbishop of Canterbury.

1011 The Danes sacked the City of Canterbury, greatly injured the Cathedral, massacred all the monks except four, and carried Archbishop Alphege prisoner to Greenwich, and after seven months' captivity killed him.

1035 Canute repaired the Cathedral and restored the body of Alphege to the Monks.

1067 Cathedral completely burnt down during the troubled times of the Norman conquest.

1089 Lanfranc, the first Norman Archbishop, entirely re-built the Cathedral and Monastery on a much grander scale.

1109 Anselm (1093-1109) the most learned and celebrated Archbishop of his time, succeeded Lanfranc, had the eastern part of his church taken down and re-erected with far greater magnificence.

1130 Prior Conrad finished the chancel and decorated it with so much splendour that it was henceforth known as the "glorious choir of Conrad."

PRICE O[illegible]

Bartle's

HANDBOOK AND GUIDE

TO

WAKEFIELD,

BY

H. C. S.

PARISH CHURCH.

JOSEPH BARTLE,
THE CITY PRINTING AND STATIONERY WORKS,
WESTGATE, WAKEFIELD.
LONDON: W. NICHOLSON & SONS, 20, WARWICK SQUARE. E.C.

Lightning Source UK Ltd.
Milton Keynes UK
UKHW031520221222
414324UK00009B/759